ROUTLEDGE LIBRARY E
JAPAN

THE DEATH OF OLD YOKOHAMA

THE DEATH OF OLD YOKOHAMA

In the Great Japanese Earthquake of September 1, 1923

OTIS MANCHESTER POOLE

Volume 41

Routledge
Taylor & Francis Group

LONDON AND NEW YORK

First published in 1968

This edition first published in 2011
by Routledge
2 Park Square, Milton Park, Abingdon, Oxon, OX14 4RN

Simultaneously published in the USA and Canada
by Routledge
711 Third Avenue, New York, NY 10017

Routledge is an imprint of the Taylor & Francis Group, an informa business

© 1968 George Allen and Unwin Ltd

First issued in paperback 2013

British Library Cataloguing in Publication Data
A catalogue record for this book is available from the British Library

ISBN 13: 978-0-415-58866-9 (hbk)
eISBN 13: 978-0-203-84510-3 (ebk)
ISBN 13: 978-0-415-84694-3 (pbk)

Publisher's Note
The publisher has gone to great lengths to ensure the quality of this reprint but
points out that some imperfections in the original copies may be apparent.

Disclaimer
The publisher has made every effort to trace copyright holders and would
welcome correspondence from those they have been unable to trace.

THE DEATH OF OLD YOKOHAMA

in the
Great Japanese Earthquake
of September 1, 1923

OTIS MANCHESTER POOLE

London
GEORGE ALLEN AND UNWIN LTD
RUSKIN HOUSE   MUSEUM STREET

PRINTED IN GREAT BRITAIN
BY WILLMER BROTHERS LIMITED
BIRKENHEAD

To
Old Friends

FOREWORD

This narrative, jotted down in the first few weeks following the catastrophe of 1923, has lain untouched for over forty years until at last transcribed in the quiet of retirement.

It was intended originally for friends familiar with the Yokohama that had so tragically vanished and with the people who dwelt there. For them no introduction was needed. But with the passage of time few can now recall the place as it used to be. For others, it may be helpful to explain how it came about that what was almost a foreign enclave existed on the shores of Tokyo Bay, and to give some description of its life and setting.

The narrative itself has been left essentially as set down at the time; and the author can only hope that the intervening years may have softened the sad memories of those who suffered grievously.

O. M. POOLE,
'Missing Acres', R.F.D.3.
Charlottesville,
Virginia, U.S.A.

October, 1966

CONTENTS

FOREWORD   *page* ix
I   Yokohama before the Catastrophe   17
II   Saturday, September 1st   29
III   Sunday, September 2nd   78
IV   Exodus   104
V   Phoenix   116
VI   Aftermath   129
Index   133

MAPS

1.   Yokohama, 1923   xiv
2.   The Japanese Earthquake of 1923   xv

ILLUSTRATIONS

2	View from the Hundred Steps—1865 American and British Consulates—1865 The Hundred Steps—1895	facing page xvi
3	Water Street—1890 No. 68 Bluff in 1917 The railing guarding the brink of the cliff	17
4	Family group in Kobe—1824 The Yacht *Daimyo*	32
5	Dodwell & Co. Ltd's foreign staff Ruins of Dodwell & Co. Ltd's offices Ruins of Dodwells' godown	33
6	The foreign settlement after the quake The upper end of Main Street	48
7	The rubble of China Town Road fissures beside the inner Creek	49
8	View across the Creek View from Camp Hill Bridge Profile of Sengen-yama	between pages 64–5
9	Homes that escaped the fire A. P. Scott's 'pancaked' house on the Bluff The author's residence: all that was left	64–5
10	Bluff Road after the earthquake Escape route down the cliffs	64–5
11	Ruins of Grand Hotel Walls of the Yokohama United Club Patched up pier and shell of Harbour Office	64–5
12	Canals and Docks were filled with corpses Bodies in the Yokohama Specie Bank Fire breaking out in the Japanese city	facing page 80
13	A scene of horror What happened to country roads Temporary graves in British Consular Grounds	81
14	A burial party in the new foreign cemetery Remains of 'Temple Court' Gate to the Customs compound	104
15	The come-back	105

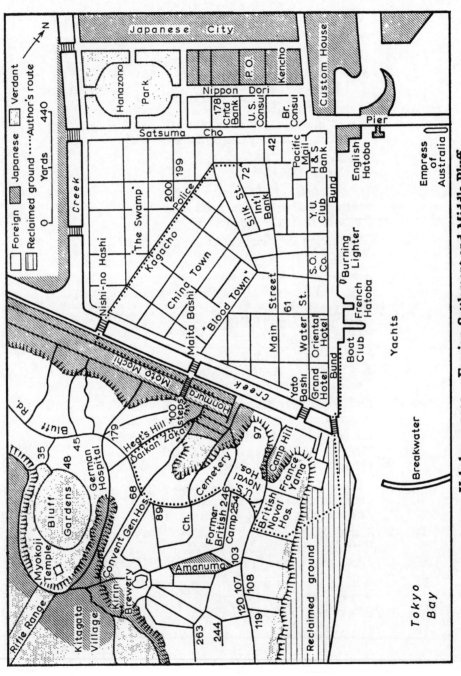

I. Yokohama 1923   Foreign Settlement and Middle Bluff

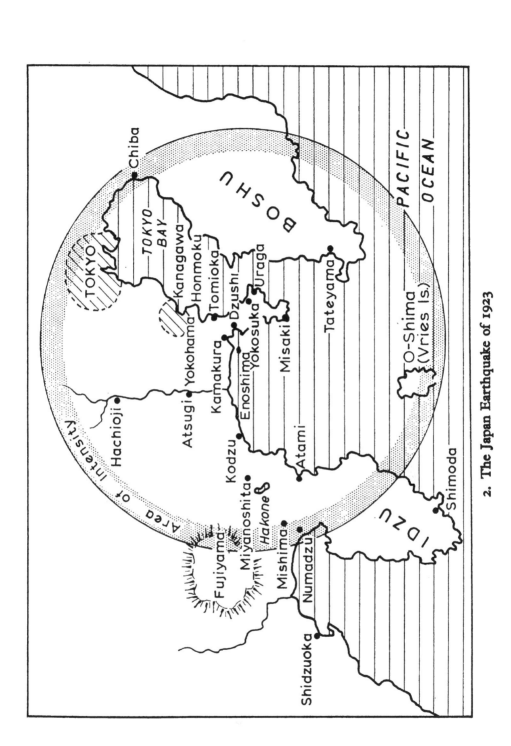

2. The Japan Earthquake of 1923

2. *Yokohama, in* 1865

View from the Hundred Steps, with Kanagawa in far distance. In the middle distance, left, are the hills of Nogeyama, below which can be seen the greensward of the future Park and athletic field. At bottom of picture runs the Creek.

in 1865

The first American and British Consulates on either side of Water Street at the Creekside. The most distant trees beyond the apex of Water Street undoubtedly mark the spot where Commodore Perry conducted his treaty negotiations in 1854.

about 1895

The Hundred Steps, leading from Honmura to Sengenyama, a spur of the Bluff. The shrine and teahouses on top were famous. Photo taken from Maida-bashi, the second bridge over the creek, leading to Honmura Road and China-town. The cross-street is Moto-machi.

3. *about* 1890 Water Street, looking north from Lot No. 32, with Mourilyan Heiman & Co's offices and tea-firing premises on left.

No. 68 Bluff in 1917. Residence of the writer and his family at the time of the earthquake.

The railing guarding the brink of the cliff in the British Naval Hospital grounds on the Bluff. Dorothy Campbell, in August 1914, standing at the very spot we had to go over the cliff in 1923. Beginning of reclamation visible below. 'Mandarin Bluff' (Juniten) in the distance.

❧ I ☙
YOKOHAMA BEFORE THE
CATASTROPHE

For over two hundred years Japan had resolutely remained a Hermit Kingdom, until in 1853 and 1854 Commodore Matthew Perry, in command of American warships, entered Yedo (Tokyo) Bay and secured the Shogun's agreement to open two ports to American ships and permit the presence of Consuls. These ports were Shimoda, on Idzu Peninsula, and Hakodate in the Northern Island of Yezo, a much needed refuge for whalers.

In 1855, Congress appointed Townsend Harris Consul-General at Shimoda and Col. Elisha E. Rice, of Augusta, Maine, Consul at Hakodate. Col. Rice arrived at his post in 1856 and held it for twenty years. It so happens that his great granddaughter is the present writer's wife.

On reaching Shimoda in 1856, Consul-General Harris, later American Minister in Yedo, renewed negotiations with the Shogun resulting in another Treaty in 1858 opening three more ports—Kanagawa in Tokyo Bay, Hyogo (Kobe), and Nagasaki in the Southern Island of Kyushu. Other nations speedily concluded similar Treaties and in 1859-60 foreign merchants hastened to establish themselves in this virgin field.

Yedo, the Shogun's capital at the head of the Bay, presented the richest prospect, but since its waters were too shallow for large vessels, Kanagawa, a populous village sixteen miles down the Bay on the Western shore, had been designated its treaty port. Though possessing excellent inns befitting an important staging post on the Tokaido—the ancient highway between Yedo and the Mikado's hereditary capital Kyoto—Kanagawa offered little scope for development, being hedged in between coastal hills and a great marsh extending two miles down the bay to the next headland. This marsh merged into paddy-fields reaching to the inner hills, a jutting shoulder of which, called Nogeyama, almost cut the marsh in two. The half below Nogeyama was contained on the seaward side by a long shingly beach backed by solid ground increasing in depth towards the headland. This strand, known as 'Yoko-hama' or 'cross-beach', had been converted to a virtual island by a creek alongside the headland connecting the Bay with the inner lagoons.

When Commodore Perry anchored off Kanagawa in 1854, he discovered deep water close to Yokohama beach, and, wishing to keep within protective range of his guns, elected to conduct treaty negotiations at an inviting spot half-way along this strand. The diaries and notes of Perry and his officers describe the lower half of the islet as 'a plain with terraced fields', which suggests that the 'little town' or 'village' of 'Yokuhama' (as they spelled it) was then congregated in the upper half. There was, however, another village part way up the creek and tight against the hillside, called 'Hon-mura' (Original Village) whose Zotoku-in temple looked down the one long street 'Moto-machi', (Ancient Road or Town). Honmura still exists, though better known now as just Motomachi.

This was the scene that met the eyes of pioneer merchants arriving at Kanagawa who, perceiving at once the advantages of Yokohama over Kanagawa as a potential trading port, diverted their activities in that direction. Despite the qualms of treaty-mindful Consuls and with willing co-operation from Japanese officialdom—doubtless relieved to see foreigners further removed from occasionally hostile elements travelling the Tokaido—work was begun on clearing the ground, bunding the water-front, improving the canals and reclaiming marshy areas. The lower half of the islet, an area roughly half a mile square, was to be occupied exclusively by the projected foreign settlement; while the upper half was reserved for corresponding development by the Japanese. No barrier divided the two sections; merely a broad avenue, Nippon-dori, starting from the site of Commodore Perry's Treaty House and ending in a patch of swamp which, when eventually reclaimed, became the 'Koyen', a beautiful park arbored with cherry trees. For the next fifty years, this park harboured the foreign athletic field, famed for its matchless turf and lovely setting.

With amazing speed, plans were drawn, streets laid out, lots allocated and building begun. As might be expected, plots along the water-front boulevard, called 'The Bund', fell to the wealthiest and most important Far Eastern firms who erected fine residential offices set back in trim gardens, with wide verandahs and living quarters upstairs for the manager and his family. All necessary 'godowns' (warehouses) were at the rear along the next road, Water Street.

Though for a while the American and British Consuls clung punctiliously to Kanagawa, expediency soon compelled a transfer to Yokohama where both established residential consulates on the creekside at the end of Water Street. In time they again moved to more imposing quarters on Nippon-dori, opposite Japanese Government edifices, where they were joined by the Russian and several other Consulates, giving Nippon-dori a distinctly official character.

Hotels evolved from simple beginnings, the ultimate 'Grand' on the creekside corner of the Bund achieving fame for its open-air terrace where travellers could bask and enjoy a sparkling panorama of tall junks, fishing boats, yachts, full-rigged ships and men-o-war passing to and fro over the blue waters of the bay.

The Yokohama United Club, higher up the Bund, provided an elite haven for male members of the community, who graciously permitted one 'Ladies Day' a year, ending in a dance.

Halfway down the Bund projected a stone jetty, the 'French Hatoba', landing place for sailors and marines from French warships. In its lee sheltered the Pavilion and Boat House of the rowing and sailing clubs, whose swimming barge and yacht moorings lay close in-shore.

Inward from the Bund, the Settlement presented an amiably haphazard array of offices, banks, fenced-in dwellings, stone godowns, tea-firing premises exuding a dreamy fragrance, churches, restaurants and well-equipped livery stables, one of which ran stage-coaches to Tokyo until the first railway was built. The chief foreign stores were grouped centrally along Main Street, beyond which the regular pattern of the Settlement was broken by an angular section known as Chinatown. Completely Oriental in character, this was a town within a town, garish, aromatic and mysterious. Around its edges, between fruit and fish stalls, sprang up taverns, grog-shops, pool rooms and low dives run by shady characters. These were the haunt of beachcombers and carousing seamen whose savage brawls made the neighbourhood notorious as 'Blood Town'. It was in one of these saloons that Rudyard Kipling first heard the yarn spun in his 'Rhyme of the Three Sealers', by no means a mythical tale. The writer himself knew two of the participants in that battle.

Last to be reclaimed from the marsh was the area between Chinatown and the encircling canal; and even after being built over with brick offices and godowns, was still alluded to by old-timers as 'The Swamp'.

The first few years of the early settlers were turbulent and not without occasional bloody incidents. To protect their nationals, both Britain and France maintained small garrisons at Yokohama. The French Camp occupied the foot of the first gulley running from the Settlement up the adjoining headland, while the larger British force camped just over the crest. Thus the road running up the gully acquired the name of 'Yato-zaka' or Camp Hill. At the top of this hill, on the left, the British also established a Naval Hospital, with terraced lawns running down to the edge of cliffs overlooking the Bay.

Similarly, the United States established its Naval Hospital on the right-hand crest of Camp Hill opposite the British Camp; but beyond its medical staff no more than a small Marine guard was quartered there. This, of course was the time of the Civil War in America.

These hills beside the Settlement formed an 'L' shaped hogsback 150 to 200 feet in height, one arm of which followed the shoreline down the bay for a mile or more with a sheer cliff overhanging the water; while the other arm led inland for about two miles, overlooking the Settlement, the swamp and the ever-spreading Japanese city. Since the Japanese prefer to tuck their hamlets into flat mouths of gullies rather than on hilltops, this hogsback and its supporting spurs were still virgin, and presented an ideal area for a residential reservation. While the Settlement was being developed commercially, homes sprang up on the hilltops, soon called 'The Bluff'; at first, little Dutch bungalows with heavy tile roofs, iron fencing and perhaps an up-ended carronade as a hitching post, but presently every imaginable type of house reflecting the owner's origins. French mansions with mansard roofs; tile and plaster cottages with bright blue porches and trellissed roses; Victorian strongholds with coy corner towers; and an occasional Georgian type, though brick was not much favoured because of its vulnerability to earthquakes. Most houses, however, were of clap-board construction in a pleasing variety reminiscent of a New England country town, with the sub-tropical addition of airy verandas, both open and glassed-in; and shuttered windows that were always closed and locked at night. All had heavy tile roofs that often slid off in earthquakes or flew off like sparrows in typhoons. Servant quarters were in separate buildings, together with the kitchen which was never in the main house. The lady of the house exercised her culinary talents strictly by remote control. With few excep-tions, each home possessed an ample garden, sometimes with lawn-tennis courts, and was hidden from its neighbours and the road by head-high fences backed by flowering hedges, trees and shrubbery. To walk along the narrow Bluff road be-tween overhanging magnolias, oleanders, cherry trees, came-lias and quince, was a delight.

Behind the Bluff lay a considerable village called Kitagata, clustered at the foot of an ancient temple, Myokoji, whose great bell and deep drums reverberated mystically on Sum-mer nights. This village led on to a crescent sweep of velvet paddy-fields and a complex of further hills in which the gar-risons laid out a rifle-range where marines and sailors from the warships regularly drilled and conducted field exercises—to the intense interest of the villagers.

In every Far Eastern port a race-course is an essential part of its sporting life. A suitable site was early discovered among the hills and enthusiastically developed. In 1907 this mile-long oval hatched a popular nine-hole golf course.

In 1912, when the community lost possession of its athletic field in the park, these same hills yielded yet another desirable site for new and larger sports facilities.

Land in the Settlement and on the Bluff was not owned outright by foreigners but held under perpetual lease from the Japanese Government, which retained ownership. The upkeep of roads, utilities, police and fire protection, postal and telegraph services and harbour control were the responsibility of the Japanese authorities. In all else the foreign community was left to run its own affairs. Official matters were dealt with through the various foreign consulates, of which the British and American stood on Nippon-dori facing handsome Japanese Government buildings, set well back on the other side. At the shore end of this avenue rose the Japanese Custom House, whose enormous compound received all cargoes lightered in from vessels moored in the anchorage. Where the Customs compound met the start of the Bund, a stone jetty known as the English Hatoba not only afforded shelter for launches and lighters but constituted the official landing place for passengers and sailors coming ashore in sampans, launches and pinnaces. The ships' anchorage, being open to the Bay, was frequently swept by storms that halted all operations. In 1891, therefore, a protective breakwater was built, starting at the lower end of the Bund and sweeping around to Kanagawa, thus creating a large, safe harbour with an entrance half-way around.

Some years later, a steel pier for passenger ships was built out into the harbour from the English Hatoba, long enough to accommodate two ships on either side berthed end to end, a boon to travellers.

Diplomatic matters were, of course, handled by the Ministers and staffs of foreign Legations in Tokyo (embassies and ambassadors came later.) The foreign community of Tokyo was never cohesive. Aside from the diplomatic corps, whose members lived in or near their scattered legations, residents consisted largely of professional men, educators, missionaries and specialists engaged by the Japanese Government such as engineers, naval architects, medical men and musicians. At the outset an area of reclaimed ground near the mouth of the Sumidagawa, called 'Tsukiji' had been set apart for a foreign quarter, but because of its remote drabness few cared to remain there, preferring other scattered spots nearer the scene of their endeavours. Missionaries, too, were widely dispersed according to their fields. Merchants whose business with the Government made it essential to operate in Tokyo, mostly chose to live in Yokohama and commute daily. It was not until just before World War I, when a long-cherished dream of the great Japanese commercial families to create a modern business centre in Tokyo began to materialize, that foreign concerns, envisaging the inevitable consequences, set about extending operations to the metropolis. With fresh enterprises starting up directly in Tokyo, a new, loosely-knit, foreign business colony gradually came into being there, increasingly dissociated from Yokohama. But that was still some time ahead.

But to return to early Yokohama. When the writer first arrived there as a boy in 1888, the sounds of hammer and saw were dying away. There remained but little unoccupied land in either the Settlement or Bluff. The site of the British Garrison Camp had long since been replanted with residences. (Not so the walled French camp-site at the foot of Camp Hill, which lay fallow many years before a handsome French Consulate rose upon it.) Such building as went on sporadically was mostly on the fringes. So far as the foreigner was concerned, Yokohama had been virtually completed, and in its existent form continued without material change through the next several decades.

Meanwhile the Japanese city grew by leaps and bounds. The surrounding areas of marsh and paddy-fields were filled in and became solid town. The population swelled from a few thousand at the outset to 50,000, 200,000 and finally, at the time of the earthquake, to 500,000. The entire marsh between the Kanagawa hills and the Yokohama headland had become a continuous sea of tiled roofs. Still cramped, the Japanese drilled a tunnel under the Bluff from Motomachi to Kitagata for a tram-line to give access to the beautiful Honmoku paddy-fields beyond the Bluff, where one could shoot snipe, quail and duck in the early morning. Into them, inexorably, the town began to creep.

In a concurrent endeavour to create more space, this time for warehousing cargoes, the authorities started reclaiming the foreshore along the base of the Bluff cliffs where a natural sandy beach, known as 'Dare's Beach', about 200 feet wide, had stretched from the Creek to Juniten where the cliffs ended. In the early days this had been a handy beach for an early morning dip, but the wash from the city's canals had begun to taint the beach so that its loss was merely scenic. The reclamation extended out about 400 feet and, at the time of the earthquake, had not yet been built over.

The foreign population of Yokohama, after reaching 2,400 of all nationalities, excluding Chinese, remained static. About 1,600 were British and American, the rest Continentals, with a sprinkling of Russians and East Indians. Since leading Japanese citizens and officials resided in the hills on the other side of the city, there was little opportunity to mingle socially, beyond formal affairs. Japanese staffs of foreign firms came to work daily from the outside, much as the foreign staffs came down from the Bluff. The faithful servants in foreign residences had their own separate quarters. Foreigners, therefore, lived a separate existence and, being all in the same boat, pulled together harmoniously. Commericial rivalries existed, of course, but were never allowed to mar good fellowship in Clubs or social gatherings, where the unwritten rule was 'Don't talk shop!'.

In the old-world atmosphere of this environment, life became singularly easy-going and tranquil. Home offices having hand-picked their men for service abroad, an unusual degree of congeniality existed. Everyone knew everyone, intimately or casually according to circumstance. Being, in a sense, isolated from the world, the community had to look within itself for entertainment and recreation. Visits of professional performers were rare. Clubs existed for every purpose. Without extravagance, one could engage in cricket, football, baseball, athletics, sailing, rowing and swimming, tennis, golf and horse-racing; and if one enjoyed shooting, snipe, quail, duck, woodcock and pheasant could be found in abundance with a little exercise. The Bluff Gardens, famed for its banks of azaleas and cherry trees, not only possessed rolling lawns and a bandstand for fêtes but a dozen tennis courts run with social felicity by the Ladies' Lawn Tennis Club.

The importance of keeping fit was recognized and office hours allowed plenty of time for recreation, running normally from 9 a.m. to noon and 2 p.m. to 5 p.m. Menfolk were thus able to go home for lunch, via the United Club for the day's news, a cocktail and, hopefully, a bit of gossip; and in the evening get in some tennis, golf, sailing or some other sport.

Social life was unflagging: 'tiffins' (luncheons), teas, tennis and card-parties; dinners, dances and National Balls, amateur theatricals, concerts and entertainments of all sorts. In those days nearly everyone possessed some accomplishment and among so many nationalities no small amount of exceptional talent existed.

Until 1899, foreigners enjoyed Extra-territoriality—(a horrendous word shortened to 'Extrality' by tongue-weary residents)—and were under the jurisdiction only of their Consular Courts. Travel was not permitted farther than twenty-five miles inland without a permit; but this mattered little as delightful spots on the seashore or in the surrounding hills were readily accessible for weekend jaunts, yachting picnics, horseback rides and excursions by bicycle. Journeys father afield and summers at mountain resorts, especially the luxurious Fujiya Hotel at Miyanoshita in the Hakone mountains of Idzu Peninsula, were easily arranged. The Pacific Ocean was only fourteen miles away and the seaside villages of Kamakura, Dzushi and Enoshima each had some individual charm.

Heads of concerns and well-to-do residents rolled around in carriages of every description, while younger blades sported dog-carts; but stands of 'jin-riki-shas' at every corner provided handy and quick transportation for those without carriages who did not actually prefer to walk. One must remember that it was not until just before the Russo-Japanese War in 1904-5 that one of the first steam-driven automobiles was seen in Japan, nor until World War I that the new means of transportation began to take hold, opening up the little-known countryside and changing long-established ways of life. Before that, the scene that met the eye on Yokohama's sun-baked roads was a clattering succession of burnished carriages crowding jinrikishas to the gutter while indomitable pedestrians, ignoring all dangers, strode vigorously among the vehicles, jauntily flicking their canes as they swung from the Bluff into the Bund on their way to work.

The unforgettable part of life in Yokohama was the closeness and permanence of friendships. Separated from homeland and kin by thousands of miles of ocean, friends took the place of relatives, bonds of affinity often proving more enduring than those of kinship. While the community was always fluid, men being transferred by their firms from one Far Eastern port to another, some families remained in Yokohama for two or three generations. There were several good private schools, but it was the general custom—incidentally known as 'The Curse of the Far East'—when children reached their teens, to send them to relatives in the homeland to complete their education. Daughters returned as debutantes to be eagerly sought by young bachelors and soon married. Sons, on leaving school, mostly stepped out into new careers; but many followed in their fathers' footsteps and came back to the Far East. Thus the community remained young at heart, with a crust of wise old-timers, each in his way a unique character, known and beloved by all.

This was the Yokohama upon which the sun rose on September 1, 1923, when my story begins.

❧ II ❧

SATURDAY

September 1, 1923

Friday night had been oppressively hot and sultry, until at 3 a.m. came a sudden violent wind and deluge of rain, flooding the upstairs verandah of our house and rousing Dorothy and me to a wet scramble, closing the wide-flung sliding windows against the summer storm.

Like many of the foreign residences on the Bluff, ours, No. 68, was a comfortable two-storied clapboard house with glassed verandahs on both floors, overlooking a modest bit of lawn and garden. It stood halfway along the main Buff road, at the head of a spur running down to the Catholic Convent. As the garden fell away sharply to a wooded gully winding towards the concealed village of Kitagata below the Bluff Gardens, our view was wide and pleasantly verdant.

During the few hours before dawn, the storm lashed itself out; and as my wife and I sat down to breakfast with our three small boys—Tony, six; Dick, four and David, three; the clouds broke and the morning turned sunny, with a strong, hot-house breeze.

Midsummer heat still prevailed, and Friday having been a holiday, a number of residents had gone off for a long weekend up country, to which happy inspiration many of them probably owe their lives. Of those who stayed in town, some took the Saturday off; but most business houses were open as usual, including my own, Dodwell & Co. Ltd., one of the most respec-

ted old British firms in the Far East. At this time, I was General
Manager of its Japan branches at Yokohama and Kobe, with
outposts in other cities.

Our office stood on Lot No. 72-A in the middle of the
Foreign Settlement, just off Main Street, on a cross-road starting from the Hongkong & Shanghai Bank. It was a newly con-
structed, two-storied building, wooden-framed, with walls of
stone blocks overlaid with granite cement. At the back lay a
long, two storied godown (warehouse) where our varied im-
ports and exports were stored awaiting delivery to Japanese
buyers or shipment abroad. Connecting the two buildings,
overhead was our chemical laboratory.

Our staff consisted of myself, then aged forty-three; Albert
E. Bateman, the Sub-Manager handling textiles and metals,
married but without children; E. C. Jeffery, Accountant;
James A. Thomson, in charge of insurance and steamship business; Frank J. Anderson, married with a five-year-old
daughter Patsy, handling lumber business with North
America; Geo. W. Colton, silk piece goods; John P. Barnett,
a recent recruit from London Head Office and No. 2 shipping
man; W. Gordon Bell, assistant to Bateman; Lucy Fox and
Mary Martin, charming young stenographers; and about fifty
Japanese, experienced seniors, *bantos* (specialists and sales-
men) and clerks. Also a Chinese 'Compradore' (Chief Cashier)
and his retinue of godown-keepers and checkers.

As midday approached, I was seated at the desk in my private office adjoining the entrance hall with broad windows
overlooking the street. A moment earlier I had glanced at my
watch and finding it just on twelve o'clock, started to close up
preparatory to walking around to the United Club on the Bund
for the usual cheery Saturday noon gathering at the long bar—
the 'News Exchange' of our exile community. Strolling to the
window, I looked up across the way to the second-floor win-

dow of W. M. Strachan & Co's brick silk godown directly opposite, waved to J. S. Stott at his desk and crooked my elbow suggestively.

I had scarcely returned to my desk when, without warning, came the first rumbling jar of an earthquake, a sickening sway, the vicious grinding of timbers and, in a few seconds, a crescendo of turmoil as the floor began to heave and the building to lurch drunkenly. Responding to the old-timers' habit, I sprang under the arch of the doorway for safety and support. During thirty-five years in Japan I had experienced numberless earthquakes, some very ugly; but though inured to their grimness and schooled to an outward show of calm, one could never escape the dread speculation, 'How bad is this one going to be?' Each onslaught seems to thunder towards a limit of material resistance, a horizon short of which lies safety, and beyond, destruction. Often before, we had hovered on the brink for a few perilous moments ere the fury flagged; and immunity hitherto from disaster helped one to face the always unpleasant ordeal with some composure. This time, however, there never was more than a few moments' doubt; after the first seven seconds of subterranean thunder and creaking spasms, we shot right over the border line. The ground could scarcely be said to shake; it heaved, tossed and leapt under one. The walls bulged as if made of cardboard and the din became awful. I saw Jeffery, in the next doorway to mine, hanging on grimly, his face round with wonderment. One moment we were flung into our rooms, the next into the hall, as if playing peek-a-boo.

For perhaps half a minute the fabric of our surroundings held; then came disintegration. Slabs of plaster left the ceilings and fell about our ears, filling the air with a blinding, smothering fog of dust. Walls bulged, spread and sagged; pictures danced on their wires, flew out and crashed in splinters.

Desks slid about, cabinets, safes and furniture toppled, spun a moment and fell on their sides. It felt as if the floor were rising and falling beneath one's feet in billows knee high. In the hallway beside me, some of our Japanese clerks were wallowing about, falling, staggering along the walls, crawling on hands and knees in a vain endeavour to regain their balance. One little office-boy, Fukui, kept creeping between Jeffery and me along the floor, dumbly raising a blanched face to each of us in turn. And then, as if heralding the end of the world, the earth seemed to rise into the air and rock; and all around us thundered the deafening roar of cascading buildings.

How long it lasted, I don't know. It seemed an eternity; but the official record says four minutes. We just hung on for dear life and waited for doom. Dimly, through the dust, I could still distinguish Jeffery rising and falling like a sailor in a dinghy in a rough sea. I don't remember thinking about the imminence of extinction; one was just suspended in awe and a savage exultation at the sheer magnificence of the upheaval.

Then, quite suddenly, stillness. No gradual slacking off; just an abrupt and rather dreaful quiet. It was over. We relaxed our holds, drew a long breath and peered cautiously about. Slowly the dust-fog cleared, settling thick upon fallen debris and the wreckage of our rooms; and the pale light revealed most of the staff instinctively drawn together in the main entrance hall by my doorway. A few were still upstairs, including Lucy Fox and Mary Martin, both slightly cut about the arms by falling plaster.

At that moment into our midst lurched Jim Thomson, half stunned, canoned blindly off the wall and was caught by Barnett. 'I had just stepped into the street', he gasped, 'and things fell on me; but I got back into the doorway in the nick of time!' I helped Barnett steer him into my chair and was

4. Family Group in Kobe after the earthquake; Spring, 1924. Mrs Calla Campbell, O.M.P., Dorothy, W. W. Campbell. Dick, Tony, David.

W. W. Campbell's yacht, *Daimyo*, our haven in the earthquake (from a water-colour by C. B. Bernard).

5. Dodwell & Co. Ltd.'s Foreign Staff in Yokohama, 1919.
Frank J. Anderson, George W. Colton, Albert E. Bateman,
O. M. Poole, W. Gordon Bell, E. C. Jeffery, Pierre B. Pattisson.
(By 1923 Pattisson had been transferred to Hongkong and
James A. Thomson and John P. Barnett had come out from
London to join the Yokohama Staff.

Ruins of Dodwell & Co. Ltd's offices at No. 72-A Settlement.
Group in the foreground are digging into the vault. Beyond
stand the brick walls of W. M. Strachan & Co. A glimmer of the
harbour shows on the extreme right horizon.

Dodwell's godown, left, fallen forwards towards the office.

relieved to hear him insist gamely that he would be all right in a few minutes. Meanwhile, the two girls had groped their way down from upstairs and Bateman called to me that all the foreign staff were now accounted for, which was good news. Equally fortunate were our Japanese staff, all of whom had escaped, only one, Koshimura, being pretty badly cut about the head.

As yet, we had no conception of what had happened round about us outside. Though our office was a shambles internally, the building itself had remained more or less upright, thanks to the retention by our architect of the unique roof of an old godown, built in the 1860's, that had previously occupied the site. Its rafters consisted of natural tree-trunks, twisted and gnarled, fitted together like a Chinese puzzle and immensely strong. Undoubtedly we owed our lives to the architect's fascination with that relic of early craftsmanship and his belief in its fantastic strength. Although both ends of the upper floor were open to the sky and our godown had toppled across the alleyway, the teak-panelled hall in which we were assembled was so little ravaged as to convey no hint of the devastation outside.

Jeffery asked urgently 'What do we do now?' I was then standing three steps up the stairway; looking over the knot of strained faces, we could see the havoc in the other rooms. Everything smothered in debris, desks piled with plaster, walls sagging, furniture flung about and the air stifling with pungent dust. Obviously, it would take days to restore order; and I knew from past experience that this first shock was likely to be followed by others, perhaps even more severe. There was only one thing to do, and that quickly. So I gave the word for everyone to 'close up and get out as fast as possible.' We weren't through with this thing yet and could come back and pitch in when it was all over.' Who could guess that we were already looking our last at Yokohama.

As they started to disperse, Bert Bateman edged up to me and whispered, 'My God, Chester, our families!' The same thought was already clawing at me out of the murk. Turning again to Jeffery, I asked if he and the other bachelors would look after the girls, and Barnett keep an eye on Thomson, who was still doubled over, because in a few minutes Bateman, Anderson and I were going to try to get through to our wives and children on the Bluff, and would have to leave them.

Anderson nipped off upstairs for his coat and hat, and Bateman and I into our rooms to close up. I have since been told that at this moment John Barnett, brimming with earnest devotion to duty, rushed up to me with a handful of papers, shedding clouds of dust and asked 'Excuse me, Sir, but would you mind signing these letters before you go?' History does not relate my response.

As I slipped on my jacket and hat, I looked around my familiar room. My sturdy Chubb safe was closed and upright, so I stepped over and spun the combination. All the steel cabinets were face down in the debris. Above the panelled wainscot gaped bare laths and struts, all askew. What a mess to tackle! But there was nothing to be done about it now and the thought of Dorothy and the boys again overwhelmed me. Had they come through? With a hollow feeling, I recalled how our chimney between the nursery and guest room swayed in every quake and split the wall-paper. Could it have fallen and brought down the house with it?

Bert Bateman re-appeared immediately, ready to go. 'It's all right, Chester, my car's at the door and we'll be up in a jiffy!' So, with a last look at the dim figures flitting methodically about the general office, we plunged with Anderson for the front door.

Never shall I forget the shock of what met our eyes as we emerged. It was like a blow in the stomach. Directly opposite, where Strachan's three-storied godown had stood, was nothing but a mound of bricks and timbers hardly a single storey high. To the right, left and beyond were jagged open gaps where other buildings had stood in solid array. In our own block, every building but ours seemed to have collapsed or was sagging like a hammock. There was no longer any road in front of our office. Masonry, beams, roofing tiles and bricks choked it in mounds the height of a man. overlaid by the entire face of our stone building which had fallen in a neat, checker-board pattern. Right under our noses protruded a piece of khaki canvas, a corner of the raised top of Bateman's touring car, squashed so flat as to be invisible. So this was what Thomson had escaped; and if Bateman's chauffeur had been trapped in the car, he could be nothing but a pancake. It transpired that he had been enjoying the shade of our wide entrance and was unhurt.

Thinking again of Stott, to whom I had waved, my eyes raked the ruins of Strachan's godown. Nothing stirred; there was neither trace nor sound. 'Poor chap', I thought, 'he's gone!' But though forty-five Japanese women had been engulfed in that godown, incredibly Stott was the first man I saw on boarding the *Empress of Australia* at dawn next day. He declared that my invitation had saved his life. Immediately after waving to me, he had left his desk and was half-way across the courtyard to their main office when the quake struck; and by crouching under an outer staircase he had escaped harm.

The question now was how to get out of the maze of ruins. Keeping together, we three scrambled precariously to the left, instinctively starting on the familiar route homeward; but at the corner of our building, where we should again have turned left down Silk Street, we were staggered to find ourselves trapped. As far as we could see, from Siber Hegners to Strahlers was a tangle of ruins, a wall standing here and there but the street jammed solid a dozen feet deep and criss-crossed with beams, apparently impassable.

The road we were on looked more promising; at least the masses of fallen masonry were free of tangled timbers and at the far end we caught the glimmer of clear ground in the triangle by Favre Brandts clock store, where Kagacho starts. So in we plunged between the tottering, shattered sides of Siber Hegner's godown and the hanging remnants of the General Exporting Co's granite walls. As we bounded into this cluttered channel, Bateman and Anderson saw, under a doorway at the corner, an elderly foreigner in a sitting posture, buried to the waist and apparently dead. They believed he was a familiar wandering character, but could not be sure. Whoever it was, he had perished, as his remains were found there next day.

I had not seen him, as, in clambering over the treacherous ruins, my eye was caught by the blue sheen of coarse homespun on the backs of two coolies submerged deep down in the debris, like a pair of porpoises. Thinking I heard a muffled groan from the depths, I heaved away desperately at the uppermost blocks, only to realize with a shudder that nothing short of a derrick could budge them. As I stood glued to the spot with helpless compassion, suddenly the ground lurched sideways and the earthquake was on again with a rumble and roar, heaving and grinding the ruins underfoot. The jagged walls on each side began to sway, shedding new tiers of stone at each lurch, and to keep one's balance became impossible. Frank Anderson, a gesticulating figure atop a mound fifty yards ahead, shouted to me urgently 'Come on! come on!', and jumping like antelopes from rock to rock, we ran the gauntlet of crumbling walls and bouncing masonry to the open square where a knot of awestruck Japanese had also found temporary security.

It was here that the full measure of the catastrophe came home to us. What seemed most terrible was the quiet. A deathly stillness had fallen, in which the scraping of our own feet sounded ghostly. Shattered fragments of buildings rose like distorted monuments from a sea of devastation beyond belief. Over everything had already settled a thick, white dust, giving the ruins the semblance of infinite age; and through the yellow fog of dust, still in the air, a copper-coloured sun shone upon this silent havoc in sickly unreality. Not a soul but our small group was to be seen anywhere. It was as if life had been blotted out—the end of the world.

On the right of the square, the stone offices of Favre Brandt, Swiss watch and scientific instrument dealers of the early days, had caved in. Ahead, on the left, lay the compound of the old British Consular jail, once surrounded by a wooden stockade. Beyond extended a flat waste of brick, tiles and timbers, all that was left of densely populated Chinatown. It seemed to have gone down en masse; as far as one could see there was nothing but gaping walls and smothered roads. Telegraph poles were either prone or crazily held up by their own lines, and an entanglement of wires overlay the rubble. A little black figure emerged from some hole a hundred yards away, and in the silence we could hear the scuttling of his feet in the crumbled bricks as he flitted towards us, like a mouse in the wainscotting.

Normally we should have skirted Chinatown to Maidabashi, the second bridge leading to the Bluff, but that route was too choked. Only one way seemed passable, straight ahead down Kagacho to Nishi-no-hashi, the third bridge over the canal. On the right, in what early foreign settlers had known as 'The Swamp' lay a section filled more with warehouses than offices, all substantially built; and many of these buildings had fared better, though most were hollow shells. Beyond that section, on the fringe of the settlement, lay the Koyen, or Park, formerly the Foreign Athletic Field surrounded by a park of beautiful cherry trees. So without further pause, we three continued running, dodging or leaping over obstacles. Where the ground was clear one had to avoid long fissures, six inches to a foot wide, one edge higher than the other.

Just beyond the Kagacho Police Station we were pulled up short. For a hundred yards or more the road had subsided and was now a muddy lake. I was in white flannel trousers and tweed jacket and instinctively balked at plunging into the muck; but Anderson—who was one of thirty-eight survivors of the South Australian Light Horse at Gallipoli—was already churning ahead up to his thighs, taking it in his stride. So without more ado, Bateman and I plopped in after him. Emerging beyond onto dry road again, we passed a brick godown which had lost its face, on the second floor of which a woman stood crying to a coolie below, who was trying to persuade her to reach out three feet to a telegraph pole and slide down into his arms, but she was too distraught to listen. I expect she did in the end, but we could not stop. By this time we were being propelled automatically as in a nightmare, perspiration channeling our dirty faces and our minds seared with apprehension for our families, but our physical senses concentrated on instantly picking out the next foothold.

At Nishi-no-hashi bridge we had expected to turn left along the good road skirting the near side of the creek to the second bridge, Maida-bashi, but one glance showed a succession of gaps where the bunded banks had slid into the water; so instead we crossed, partly on hands and knees, the collapsed iron bridge into the narrow strip of native town called Motomachi, thinking to climb straight up the Bluff by the Cherry Mount

Hill. Once across, however, new impediments greeted us. The shoulder on which the Cherry Mount stood had avalanched down into Motomachi, carrying half the hotel with it. Where the stepped lane had climbed a steep notch, nothing remained but the slippery scar of a wide landslide. All along the fringe of the Bluff, virtually a long cliff, similar landslides had carried everything down into the village below; and we could see that most of the foreign houses near the edge had crumpled or stood tottering.

Nothing, however, could equal the appalling state of Motomachi. For its entire length, to the temple below Dr. Wheeler's, it had been flattened to an unrecognizable tangle of matchwood. The long road of gay, open-fronted native shops had disappeared, its location indicated only by a V-shaped channel of jack-straws where houses had crashed together from both sides in heaps of heavy tiles and splintered timber. At every few paces stood a tattered human figure, shocked to immobility by the engulfing of practically the whole population. One felt that somehow the multitudes of creatures under that wreckage must surely give a mighty heave and emerge alive in their hundreds. But all was silent; only here and there a shriek from the depths, or encouraging shouts from some frantically digging group.

Aghast at the scene, we looked back the other way towards Jizo-zaka and gasped. In the few moments since we had crossed the bridge, fire had burst out in Motomachi not eighty yards from where we stood and red-brown flames were spiralling into the air in a vicious whirlwind. The unnatural hot-house breeze of early morning, whipped now to a torrid summer blast, picked up masses of flame and swirled them straight towards us along the groove of ruins. In an instant the fallen woodwork became alight and new flames raced towards us like a prairie fire. At that a great sob rose from the survivors round

39

about and the anguish in that wail was so pitiful that it seemed to numb one's senses against the impact of any further tragedies to come. One became an automaton with a single purpose.

All this was happening faster than I can tell. Our only course now was to keep ahead of the flames by fleeing along Motomachi towards Maida-bashi; so down that gruesome gutter of the living and dead we stumbled together, trying not to step on the occasional arm or leg protruding from the wreckage. Here and there a knot of bewildered people huddled on some mound gazing dumbly at the approaching flames.

After about a quarter of a mile, we reached the familiar point where the famous Hundred Steps led steeply down from the Sengen-yama, a Shrine and fabled teahouse at the very edge of the Bluff, into Motomachi opposite the Second bridge. Gone were the steps and half the cliff with them, obliterating the houses below. Only a glistening scar marked the spot, and a hillock of earth. With the fire still licking at our heels, we pressed on another block and turned up into Daikan-zaka, known to old-timers as Hegt's Hill. At the corner stood an unroofed grocery store, its pyramids of canned goods still almost intact in open air. Suddenly, at the thought of food problems ahead, I hailed Bateman and together we filled our pockets with condensed milk and canned meats, the shopkeeper mechanically computing the cost on his soroban as if nothing had happened. It only amounted to a few yen, so cramming a five yen note into his hand, we dashed off, leaving him agitated over the ignored change. Probably his last sale!

Frank Anderson, missing us, had meanwhile gone on ahead; and at this point Bateman and I also separated, he to go up the Foreign Cemetery hill as the shortest route to his house at No. 107 Bluff, and I to take Daikan-zaka as the safest route to mine at No. 68, beside the General Hospital. About half-way up, twenty Japanese were watching the rescue of a touselled teahouse girl from the upper floor of a teetering tavern. I had scarcely passed when another earthquake shock struck with a rumbling roar; and looking back, nothing could be seen but dust and scrambling figures where the tavern had stood.

Now, as I approached the top of the hill, I could see close at hand how residences on the Bluff had fared, and my heart sank. To the right, the second house from the top that had once been 'drunken Lewis'' was crushed as if someone had sat on it. Above it, on the corner of the main Bluff road, No. 59, best remembered as once the hospitable home of the Hills and their two fair daughters, Winnie and Doris, had also been bashed down and was unrecognizable. Across the hollow to my right, on a spur of what was in my boyhood the playground of the Victoria Public School, a slide of bricks and timbers spilled down the slope—all that remained of a six-storey apartment house known as the Retz Building. At the top of Hegt's Hill, to the left, two of the early-day low Hegt bungalows on Lot. 73 had 'pancaked'—the uprights collapsed and the tile roof like a pie-crust over jack-straws. So also had fared two bungalows across the road. And looking down the opposite hill to the German Naval Hospital and wooded spur of the Bluff Gardens, it seemed that every house was either down or in tatters. I quailed at the completeness of widespread ruin and my fears for Dorothy and the children became a desperate prayer.

It was now only a couple of hundred yards to our house around a tree-lined curve which still concealed it from me. As I swung left into the Bluff Road, I met Hugh Sharp of the Hongkong & Shanghai Bank, who had beaten his way up along the Bund and Camp Hill, and who grasped me momentarily by the hand in passing, saying 'Your house is still standing,

Chester!' 'Thank God!' was all I could reply, not having the heart to tell him that his house, opposite the gate to the German Hospital, was badly knocked about. His young wife Joan, Dorothy's dear friend, came through but their baby boy was killed in her arms and was buried two days later at sea.

With eyes fixed forward, I ran on through broken fences until at the turn I caught a glimpse through the branches of Warrener's house, next door to ours and still upright. A moment more and there was ours on the corner, battered and tottering but not completely down. And there too, out in the road waiting for me, was our old house-boy Ishii who had been in my family for years, blood trickling from a nasty gash in his leg. 'Are you unhurt?' he called anxiously, seeing my dishevelled condition. 'Yes, yes! But Boy, how about the okusan?' '*Daijobu, mina-san daijobu!*' he assured me. (Safe, everyone safe!) 'They got out into the garden through the verandah and I think went down the lane to the convent gardens to wait for you.' Immeasurably relieved and grateful, after a few more hurried words I dashed off down the lane to resume my breathless quest.

On my left, the General Hospital hung like a hammock between its two wings. On my right, the homes of Jock Watson, Paul Blum and Ed. Mendelson were little more than frames. Farther down the lane on the left lay the convent, and I panted into its garden, shouting Doro's name; but the only response was from a black-hooded nun who toddled forward from a knot of frightened children and, pointing to the shattered buildings, begged in French for some kind of help. I longed to comply, but when I explained that I was seeking my own wife and family, she understood and waved me on my way. Obviously Dorothy could not have gone beyond the convent as the narrowing lane was completely blocked by a high wing which had collapsed outwards, so I sprang through the gates of the No. 2 Hongkong Bank house occupied by our friend, R. C. Edwards, the terraced gardens of which offered a good refuge. But look and shout as I might, there was no sign of them and as by now the sky was filled with rolling black smoke from the Japanese village of Kitagata below, I was becoming each moment more desperate. Then, as I was scrambling along a bank through a knot of servants and villagers regarding each other helplessly, came yet another terrific shock that rocked us off our feet, dislodged the bank and slid us all down the slope in a miniature avalanche, or earthen toboggan, depositing us unhurt. I wondered then, for the second time, whether all this was not just a nightmare.

It began to dawn on me that I must be on a false trail, as Doro and the children would not have hidden themselves so completely. So I ran back up the lane and there, at our corner stood Anderson, gesticulating wildly and, as I came within hearing, shouting to come on over to his bungalow only one hundred yards away on the other side of the General Hospital —No. 89, where we Pooles had lived for thirty years until Mother's death two years after my marriage. There I found Dorothy and the boys, together with Honor and Patsy Anderson, in the garden; and the relief and subdued joy of our reunion was like waking out of a bad dream. They were all sitting on the grass in the shade of a large cedar tree, in the very spot where thirty years earlier Father, Mother, Bert, Eleanor and I, with a few friends, had camped out all afternoon and evening after an earthquake, till then the worst experienced in Yokohama, which had shorn No. 89 of its tiled roof and done widespread damage. Now, however, No. 89 lay pancaked, just a heap of beams and sticks capped by the peaked roof, only a wing bathroom and corner of what used to be mother's bedroom uncollapsed.

When the quake started, Honor and Patsy tried to bolt out

of the front for the garden, but as they reached the door to the verandah, down came the house about their ears. By a miracle they found themselves under a tent of timbers, and though severely bruised were able, when the dust subsided, to worm towards a glimmer of light and freedom, emerging just as Dorothy and the boys came over from No. 68. With them, devotedly standing by their charges, were the childrens' two Japanese amahs—grizzled old Miné-san and pretty young Kané. There too, in the group when I arrived, were Dorothy's father W. W. Campbell, General Agent of the Pacific Mail Steamship Co. and Commodore of the Yacht Club, affectionately known up and down the China Coast as 'Willy-Wally'— and her equally beloved mother 'Calla', still a vocal favourite and Inter-port tennis player. 'The Commodore', laid up for a few days with an arm poisoned by red jelly-fish, was luckily at home with Calla when the earthquake struck. Their house —No. 35 down the Bluff Gardens Lane—was split from top to bottom as by a sword, and when the high stone foundations crumbled, squatted unsteadily on the ground. Finding themselves uninjured, their first thought was of Dorothy; and leaving their servants on guard, they raced along the Bluff Road half a mile to our house and on to the Andersons. That we were all safe and reunited seemed incredible after the terrible things we had seen.

Also momentarily in the garden group was the keen golfer A. T. White, who had escaped intact from the General Hospital. He helped out others and then assisted Dorothy and the children from our house to the Andersons', even recovering Tony's pith helmet from the ruins. The heat was pitiless and Dorothy ingeniously contrived hats for Dick and David out of lampshades stuffed with leaves. A neighbouring couple, the Kings of Priest Marian & Co., fleeing from their demolished home, also joined our group; and although Tiny White presently disappeared, the Kings cast their lot with us through what followed.

It was stifling hot and the rolling smoke overhead seemed to be coming lower and lower. Anderson, who had momentarily disappeared, emerged with a handful of lemonade bottles he had dug out of the icebox on the back verandah, and a gulp or two felt like nectar to our parched throats. Only then did I begin to learn what had happened to them all.

Dorothy and the children had escaped harm by providential chance. Just before noon she had discovered that David was running a 101° temperature, so with Miné's assistance, she gave him a dose of oil in the upstairs nursery. Though the youngest, he always took it like a soldier, and the other two left their game of blocks in front of the fireplace to come and watch. Thus all were close together when the earthquake struck. With one accord, Dorothy and Miné made a circle of their arms around the children just as the defective chimney fell crashing through roof and floor precisely where Tony and Dick had been playing a moment before, flinging the group to their knees but not striking them down. Still herding the frightened children, they crawled to the doorway and huddled there while the earthquake twisted and wrenched the house asunder. When all turmoil ceased, they managed to clamber down the skeleton of the staircase to the lower entrance hall. The closed front door was, however, jammed by the fallen hatrack, the door to the kitchen blocked by debris; and there was no way through the dining room because the other chimney had crashed its full length over the table.

There remained the drawing room, directly under the nursery, but the chimney was piled on one side, the piano had slid across the room, blocking one French-window, while the sofa jammed the other. However, Ishii, who had been setting the table and had made for the garden, came back into the verandah and lifted the children over the piano, Dorothy and Miné following. Finding that the terrace which supported our small lawn and shrubberies had mostly slipped down into the valley, they edged around the house to the front gate: and after standing awhile in the open road, had gone on unobserved by Ishii to the Andersons' garden for shelter under their trees, accompanied by Kané.

Although only an hour had elapsed since the earthquake first struck, it was already evident that we were to be given no respite. The billowing canopy of smoke overhead was now as black as a thunder-cloud, shedding burning chips as it swept past. With fallen houses like prepared bonfires on all sides, our danger was pressing, and flight unavoidable. Dorothy would not consider my diving back into our house to see if I could lay hands on anything that might prove useful, and indeed there was no time now for anything like that. Every minute or two the ground swayed afresh, and crashes here and there signalled continuing demolishment. Anderson drew me aside to whisper 'Can't we get out of here? I think this is something more than an earthquake; there are fissures back of the house and steam seems to be coming out of them!'

What this could have been I don't know, but anything seemed possible then. Strangely enough, I had often had such dreams in my boyhood in that very house, so I was not in the least incredulous. The Commodore, too, did not like the look of things and joined us, declaring, 'Son, the water is our only safety now. Let's all get down to the Bund and aboard the *Daimyo*. There are some provisions on board and we can live this thing through.' The Commodore was a hardy sailing man and the *Daimyo* his graceful, seven-ton racing and cruising yacht with a commodious cabin, at that moment lying off the boat house in the yacht anchorage, directly in front of the Settlement Bund. It did seem to offer the best haven; so, picking up the children in our arms, we set off with hardly a backward glance at our houses that we never saw again. The poor children were barefoot and hungry, but not a whimper came from them or from little Patsy, although the fearful havoc must have struck terror to their hearts.

The little Bluff Library at No. 91, almost opposite No. 89, had vanished down the valley. The Bluff road was like a flight of steps, fissures eight inches wide every few feet, each ledge sagging successively towards the valley opposite. All along the road and in the gardens on the right, these ledges ran from northeast to southwest, indicating a massive shift in the hog-back's structure. E. J. King's large residence of French tile and brick, of which the library was originally the gate-house, was just a heap of rubble; both his wife and daughter were killed.

Just here we met George Allcock, tall, grimy and haggard, striding towards his home below the convent, now a vortex of smoke. Grasping the Commodore's hand in passing, he exclaimed 'Thank God you're all safe!' and was gone. Tony, riding astride his grandfather's shoulders, suddenly asked 'Commodore, why did that man say "Thank God"? Why should we thank God when he does all this?' and he waved his arm toward the burning city. 'I wouldn't be as cruel as God!' A fierce reaction in a child of six! I was glad to hear later that Allcock got through to his home and family in time and they escaped safely. But six of the convent nuns and a number of the Japanese children lost their lives in the convent.

We were now at Christ Church on the next corner to 89, which looked as if it had been blasted, only one arch rising complete, the rest in massive chunks of brick and steel spread over the road like a barricade, breast high. There was nothing for it but to clamber across, and as we did so another quake ground the blocks against each other like trotting circus horses rubbing shoulders; but luckily, we kept our balance and were unhurt. 'How strange', I thought, 'this is the second time I have climbed over the ruins of this church!' Twenty years earlier, while it was being built and only the light steel reinforcing frame was up, J. H. C. Goodban and I, both in Dodwell & Co's shipping department, were fighting our way at 5 a.m. along the Bluff to the harbour during a typhoon, when the swaying structure careened over our heads and we raced from under just ahead of the crash.

From the church we could now look over and beyond the Foreign Cemetery, which occupied the side of the Bluff ridge sloping into a valley at the head of Motomachi, and see practically the entire Settlement and Japanese city stretching to the Nogeyama hills. But there was no Settlement or city; nothing but a sheet of flame and smoke. The fires that started in a hundred places, wherever buildings collapsed over hibachi or stoves, had spread unchecked in the violent wind, and within an hour all Yokohama was simultaneously aflame. Motomachi, through which we had run, was already consumed up to the temples below Dr. Wheeler's, and flames were licking up towards his house through the trees. As for the cemetery, it was in tragic state, the black surrounding fence flat, tombstones flung about and broken, and some of the graves yawning where fissures had split the small terraces. The granite Memorial Arch to those killed in the Great War lay in a dozen big fragments. That even the resting place of the dead should be heaved into such disorder seemed like pure sadism.

Opposite the cemetery gate, Carew's house—the old Merriman place of happy memories—was standing in surprisingly good shape; but the American Naval Hospital on the left had come down badly, and the bunded wall of its wide garden, extending from the cemetery to Camp Hill, had fallen into the road, carrying some trees with it. Tim Brady's house, beyond

6. Foreign Settlement of Yokohama two or three days after the quake. Stretch of Creek above Maida-bashi, right. Road in foreground is Motomachi. At centre, coal piles still burning, Yokohama Engine & Iron Works at left. Yokohama Harbour in distance, with partially submerged pier at left.

Foreign Settlement; upper end of Main Street. Gutted Russo-Asiatic Bank, right. Dodwells' offices just out of picture, left. Roads half-opened two weeks after the quake.

7. The Foreign Settlement. Rubble of China-town at the bend in Honmura Road. Tall ruins in background mark Singleton Benda & Co.'s offices at No. 96.

Fissures in road beside the inner Creek. Foreign Settlement: 'The Swamp' area, left. Crowded Japanese city, right. Residential Bluff in distance.

the Carews, had pitched forward into the road over the debris of the hospital garden, blocking it completely and forcing us to detour over the hospital lawns to regain the road at the top of Camp Hill, which ran down to Yatobashi, the first bridge at the mouth of the canal, where it entered the harbour at the corner of the Settlement by the Grand Hotel. At that point the Bluff road took a sharp turn right, following the line of cliffs dropping down into Tokyo Bay; and on the corner facing Camp Hill stood the Gaiety Theatre, a red brick relic of earliest days but still the scene of all our theatricals, balls, concerts and public meetings. Where it had stood there was nothing. It had disintegrated and its masonry been completely swallowed up in its cavernous cellar. One looked down a smooth valley to the Japanese village of Amanuma in a fold of the foreign bluff. Just so must it have appeared in 1860 when this valley was the British military camp.

It had been our plan to get down Camp Hill to the Bund and so aboard the *Daimyo,* but here at the top of the hill our trek ended. The native village which filled the gully up which Camp Hill wound, was a roaring furnace, and the road partly choked by a landslide from the old E. J. Moss house. We pushed a little way down, trying to find a way through the French Consular residence on Fransu-yama, the last spur of the Bluff: but the stone-bunded wall of the British Naval Hospital, and the ward on top of it, had collapsed across the entrance drive, closing it off. To push through the woods on the steep side of the spur would have been hazardous even without the landslides and hot smoke, and there was no saying that we could get past the fallen residence if we did.

So here we were, penned on the Bluff, with the fire coming up from behind.

Up to this point the two amahs had loyally stayed with the children, but Dorothy suddenly thought of Kané's twelve-year-old son who went daily to school in Kitagata below the convent. 'Kané' she exclaimed, 'what of your boy? Don't you want to go to him?' 'Oh, yes,' Kané faltered, 'if you can spare me!' So, after a quick farewell, she headed back with one or two companions along the way we had come. Weeks later we learned they could not get beyond the cemetery because of the approaching fire, and rode through the conflagration, crouched behind tombstones with kimonos around their heads, almost suffocated, the scorched grass burning round about. Old Miné, whose people were simple fisher-folk at Sugita, refused to leave us. Her dignity and courage were wonderful.

It was no situation in which to linger on the dusty road among broken glass, with the barefooted children already suffering from the overpowering heat. Now that we could not reach the *Daimyo,* some other haven was essential, and we turned instinctively to the British Naval Hospital beside us; and climbing its steps through the Torii gateway, we entered its spacious grounds. To our amazement, every ward and bungalow had been laid flat, but scores of people had already taken refuge on the double tennis-court above the Bluff road where we had so often played, and others were making their way down to the next terrace towards the cliff overlooking the bay, and thence to a third terrace at the very edge. Here, however, the lawn was alarmingly fissured and one end had already fallen away to the reclaimed ground on what had been a beach below the cliffs, until a few years ago. Every few minutes a low rumble and dizzy lurching of the ground warned that we were still in the midst of an upheaval, and in fear lest the whole third terrace might slide down the 200-foot cliff, we threw ourselves down under the shade of some deodars on the next terrace above, feeling safe at last.

For the first time one could pause to take stock of the situation. If we were to bivouac there, we should presently need some kind of shelter, food and water; so, assuring the others we would not be gone long, Frank and I went back up among the fallen wards to see what we could find. Unfortunately, the ruins were so tangled that an armful of pillows and sheets was all we could secure. The sheets would make tents and for that we were grateful. But the only food we had was what I had dumped from my pockets. Here we came across T. M. Macgregor, the hospital's accountant, scarred and shaken through having been buried for over an hour, but otherwise all right. And as we started back with our loot, beside the ruins of his bungalow stood Dr. Hingston, his cheek torn and bleeding, and numbed with grief at discovering that his wife, still deep in the timbers, was dead. A couple of British sailors were trying to extricate her and the doctor asked if we would help. After a bit, I went down to let Honor and Dorothy know what was keeping us, returning just as they were lifting Mrs. Hingston out of the hole, so pathetically young and supple that it was hard to believe her dead. They laid her on the grass and we did not pass that way again. Dr. Hingston was wonderful after that.

Returning to our group under the trees, we found them joined by Mrs. Coutts, with big Edward Coutts on a stretcher beside her, his back injured by the fall of their staircase upon him. Fortunately, they lived only a block from the hospital at No. 112, and she had been able to get a stretcher party to move him. Alongside lay the injured Russian nurse-girl of the Stapletons who lived at No. 110, opposite the Coutts. I spoke to Coutts but he was in great pain.

Just then it suddenly occurred to me that we had seen nothing of Bateman or his wife Gladys, and I felt a sharp presentment that he was in trouble. Seeing C. E. D. Parkhouse of Samuel's standing nearby, I asked if he would accompany me on a sally to find Bateman, whose house, No. 107, lay only a few hundred yards along the Bluff road. He consented at once; so, with a word to Dorothy, we dashed off. Maurice Russell's large house next the hospital grounds was standing practically intact, but the Selby's opposite, beyond the Gaiety, had gone completely, as had Dell Oro's. Morris Mendelson's house, No. 103, at the corner above the hill down to Amanuma village, was still upright though distorted. Concerned for his pretty bride Madeleine—I had attended their wedding in London the previous Christmas—I nipped up the garden path and into the house, shouting her name. There was no response and obviously the house was empty. Poor girl, she was caught downtown, shopping in Brett's Pharmacy, and she and their chauffeur were overwhelmed while getting into the car and burned to death.

We were now at Bateman's house, the second beyond the corner, which was leaning outwards over the street, prevented from falling only by a stout telegraph pole. Edging warily through the door into the disordered hall, we called lustily, but aroused only hollow echoes. By this time the fire was approaching fast up through Amanuma village. An idea struck me. 'Quick, Parky, over the way! They may be in the Syme-Thomson garden!' The house, No. 108, directly across the road, was completely down, only the kitchen standing and the rest swirled down a drop into the lawn beyond. Jogging through the rock garden, we could see four or five persons clustered, one of whom detached himself and, for all the world like a black Gollywog, staggered towards me with arms lifted high over his head, and flinging them around my shoulders almost sobbed. 'Oh, Chester, I've only just dug them out, safe! Not till I heard his voice could I recognize Bateman. And then I discerned that two limp figures in wicker deck-chairs with towels over their heads, were Mrs. Syme Thomson and

Gladys Bateman, tattered and only half conscious. The other two were Sada, the house-boy, and the gardener.

It seems that Gladys had come across to sew with Mrs Thomson and they were sitting together on the sofa when the first convulsion somehow flung them between the sofa and wall in such a way that the falling timbers could pin but not crush them. When Sada, who escaped from the kitchen, returned to look for them, he did not know where to begin digging until Mrs Thomson's tiny Schipperke started clawing and barking down into the ruins, from which presently came feeble cries. At this point Bateman reached them and the three began frantically to dig and heave. Half-way down a great beam, lying athwart other timbers, blocked further progress, defying their united strength. They started despairingly to look for a saw and found a little pruning saw in the gardener's lodge. With that, taking turns, they managed to saw through the beam in an hour, and tearing away the remaining few feet, were able to lift the prisoners free. Painfully bruised and exhausted, it was not until next day they found that each had a rib or two broken, but were otherwise uninjured.

It was then that Parkhouse and I arrived on the scene—and only just in the nick of time. The smoke sweeping across the road was turning red and hot, demonstrating the necessity of getting everyone out of this pocket fire-trap without an instant's delay. We tried at first to carry the women as they were, in their chairs, but the going proved too rough over the ruins and they preferred to stumble along on foot with such support as was not too painful. Once on the road it was better going, though the fire breathed hotly on our left cheeks. Easing them along gently, we presently had them safely in our camp under the deodars. Not a moment too soon, for looking back we could see flames lashing at Bateman's house and arching over the road.

By twos and threes the number of refugees was swelling, most of them wandering aimlessly on the upper tennis courts. Some were dishevelled, some bandaged, and many dazed. A few were Russians, refugees from Siberia who had settled in Yokohama temporarily; and here they were again being cruelly uprooted. Over a score were killed in the collapse of the Retz Apartments. One fair-haired girl of seventeen, clad only in a nightgown and kimono, was standing on a grass hummock beside the path so bewildered as to be unaware that her torn garment was revealing a lovely young figure. Against the swirling black smoke she might have been a Grecian statue.

As the encircling fire grew nearer, we made further sallies to reconnoitre. Anderson and I probed again as far as Morris Mendelson's house and made sure it was still empty. Beyond the hill, down to Amanuma, the fire had already crossed the Bluff road and was sweeping down to the cliff-tops. We were now cut off on that side. Returning to report, we found that H. W. Rowbottom of Samuel's, alarmed at discovering that the exit lane from the hospital grounds to the creek was blocked by fire on Fransu-yama, and by no means convinced of our safety, had enlisted the help of three British sailors in stripping the tennis courts of their surrounding nets and dragging them down the terraces to the edge of the cliffs. Making fast one end to a summerhouse overlooking the Bay, they tossed the rest over the cliff where it half dangled, half rested against the grass-tufted face. I went down to have a look over the edge. It was a precipitous but not perpendicular drop of 150 to 200 feet to the beach and reclaimed ground and the net seemed to reach two-thirds of the way down, disappearing over a ledge where it had apparently to be abandoned, as two or three figures, like flies on a wall, were picking their way horizontally along a narrow lip for about fifty yards to a fresh landslide down whose brown slope they could descend to the flat below. Rowbottom was urging everyone who felt up to it to make the descent while the going was good, but after one timorous look there were no takers, so he came over to join me.

He said his wife Flora and her father Henri L. Fardel, my old Swiss school-master of Victoria Public School days, had left Samuel's office at No. 27 Settlement in their car ten minutes before the 'quake and he had no idea what had happened to them. Fires had frustrated his attempt to get to his children at 225 Bluff, almost a mile beyond our house, and had driven him back to the Naval Hospital. Now all he could do was to help others and pray that his family would turn up safely. Days later, the car was found in Main Street, half buried, with Fardel's charred body relaxed in the back seat, sitting where he must have been killed by falling masonry before the fire reached him. Flora's body was found later under the rubble on some steps nearby. The children came through safely.

The Commodore meanwhile had joined me and together we studied Rowbottom's escape route, concluding that while an active man with a good head could chance it, taking children and injured women down would be too hazardous. So we decided to stick it out on the terraces and hope that the fire would pass. This it seemed to be already doing on one side, Maurice Russell's residence having caught and kindled the next below it, so that now the whole dip from the lane skirting the hospital grounds to the Syme Thomsons and from the Bluff Road to the cliff-top was a mass of flame roaring out over the bay. So far, the houses directly across the Bluff Road from the hospital had not caught, nor had the Gaiety, Brady's, Carew's or P. B. Brown's opposite the cemetery gate; and our women-folk told us that Bateman and Anderson had gone off to the latter group to scavenge for food in the ruins.

In spite of a screen of tall trees, the noise and heat from

the river of fire across the side lane, only a hundred yards away, was becoming rather frightening, and the Commodore and I slipped away on another reconnaisance down Camp Hill, on the chance that some way of getting through had opened up; but it was still a cauldron of flame. On a square of matting in the middle of the road, as close to the fire as he could get, knelt a solitary old Japanese, bowed in dumb prayer. Probably his family was in that inferno.

A tip-tilted sedan car had been abandoned nearby, two wheels trapped in a fissure; and a little beyond lay a corpse, apparently pulled from a landslide that partly blocked the curving road.

Seeing there was still no way through to the water, we retreated to the top of the hill where, with a shock of alarm, we found the situation had undergone a startling change in the last few minutes. From somewhere below the Gaiety, the fire was now surging directly for the hospital grounds and was already lapping at the houses along the Bluff Road. When they went, nothing could stop the flames hurdling the road to the wreckage of the hospital wards strewn on the ground like prepared bonfires. 'My God, Commodore, we're in for it!' I exclaimed. 'We'll have to go over the cliff now, whether we like it or not.' 'Looks that way, son' he agreed, 'and no time to waste.'

Hastening back to our party, we were dismayed to discover that Bateman and Anderson had not yet returned from their sally; and their wives and Mrs Syme Thomson, with white faces, refused to budge without them, fearful lest they lose each other in the confusion. The only thing to do was to try and find them, which Willy-Wally and I essayed in another quick sally; but it was like looking for a needle in a haystack in all that smoke, and we beat our way back, hoping they would turn up any moment. We did not know till later that they had gone as far as Percy Brown's house and, engrossed in groping between its walls, were unaware that the fire was cutting in behind them.

People were now beginning to leave the upper terrace and gravitate towards the cliff-top. As soon as they started to use the net rope, our problem of getting the children down would be infinitely greater. The time for action had come. Assured that Anderson and Bateman would appear any moment, I decided to make a start with our small boys right away, realizing that I should have to make three trips of it since the Commodore's poisoned arm made it impossible for him to risk carrying one. Patsy Anderson would naturally stay with her mother and come with Frank when he turned up. This quickly settled, Mr and Mrs Campbell, Dorothy, the three boys, their amah Miné, and I moved down to the summer-house and told Rowbottom we were going down.

Tony being the heaviest, I was to take him first, and the sailors soon had him bound to my back with a short length of rope. They then made a noose at the end of a long one and passed it over his shoulders so that they could pay it out as I descended and eased the strain on me. Dorothy went over first in order to be able to take Tony from me when we reached the avalanche, and in trepidation I saw her dangle and swing down from foothold to foothold. 'Hold tight!' I called anxiously. 'I'm all right,' she answered cheerily. 'I climbed trees as a child and I can do this now.' Giving her a twenty-foot start, I followed, the Commodore keeping Tony's rope taut, I had climbed around those cliffs as a boy and there was something almost familiar about doing it again in these fantastic circumstances. All went well for the first twenty or thirty feet, but as the cliff face grew steeper, the descent became tougher and, being out of sight of those above, they could not tell how fast to pay out Tony's rope and it began to cut into his shoulders.

The brave lad had showed no fear at going over the cliff but now the pain of the rope made him cry out and I had to swing him off my back to my chest, where I could partly support him with one arm and tug at the rope for more slack as we dropped from one foothold to another. Bit by bit we worked down to the end of the tennis net where W. King had stationed himself to help people off the line and point out the way along the perilous ledge to the landslide. To negotiate it called for nerve, as it was little more than two feet wide, irregularly humped, while immediately below the cliff was undercut and dropped invisible, fifty feet to the beach. A slip would be fatal. Used to mountain climbing, I should not have minded it alone, but I feared for Dorothy and the children. However, she scrambled across as nimbly as a cat; and at the landslide I handed over Tony and saw them start down in a sitting posture, side by side.

Dorothy reminds me of an amusing incident here. As I left her and Tony at the top of the slide, a Russian woman, glued to the spot, was wringing her hands with fear at having to descend the steep slope. Dorothy knew no Russian and for a moment only Japanese words would come to mind, until suddenly out poured a mixture of German and French: *Es macht nichts, Madame, il faut glisser!* Whereupon a wan smile lit the woman's face and she followed Dorothy and Tony down the slide on her plump bottom.

Sidling back over the ledge, I climbed almost hand over hand up the net-rope to the cliff-top. There they all were, Miné the amah with her beloved David already strapped to her back, set on descending with him so that I might take Dick and all of us go together. This would have been folly; so, whisking him off her shoulders, I had him roped to mine, eased over the edge again and so on down. The others were to await my return.

This time it was rougher going, as those following us were dislodging clots of earth which bombarded and sometimes struck us, luckily without harm, We made the traverse safely, although in one place spring water was oozing from the cliff face, creating a soggy gutter across the ledge which grew deeper and wider as each person slithered in it. Dorothy and Tony were waiting at the bottom of the landslide and I planted them further out, beyond range of possible further slides, before leaving them again.

As I started the upward climb once more, I saw at once that what we feared had begun to happen. Many others up above had realized the growing danger and were coming down fast. The Commodore, perceiving this, had made Mrs Campbell and the amah take the plunge without him and they passed me on the landslide, downward bound. As the throng on top increased, Commodore decided I could never get back up the rope against the stream, so ignoring his disabled arm, went over the cliff with Dick roped to his back. Being a hardy yachtsman and strong as a bull, he was getting along all right until Dick's lashing came loose; from then on, he could only press Dick to him with his game arm, while gripping the net with the sound one.

I had just begun to cross the ledge when I heard Doro's anguished call 'Hurry, Chester, and help father; he's in trouble with Dick!' I caught a glimpse of him then, nearing the end of the rope with Dick falling outwards from his back, saved only by two of his small fingers that were hooked to the Commodore's collar at arm's length. Reaching them in the nick of time, I snatched Dick from his predicament and, with his arms around my neck, I wormed cautiously back along the ledge. Now, however, a gap had developed where the spring-water was carving a chute to the beach below. To jump it with Dick in my arms was too risky, so, setting him down, I sprang across, turned about, kicked a foothold two feet down the gap and, holding out my arms, commanded him to jump. Without a second's hesitation, he came through the air like a star-fish, landing against my chest with the grip of an octopus. Gingerly, I regained the ledge and in a few minutes we were all reunited on the beach below.

There I ran into H. T. Stapleton of the Chartered Bank, black as a sweep with smoke and grime, with whom I was to have played golf at 2 p.m. 'Nice game of golf we're having!' I quipped wryly, then stopped short at the sight of his drawn face. He had not found his wife and children yet, they were cut off by fire. (They were saved; I forget how.)

In growing alarm at the non-appearance of the Andersons, Batemans and Mrs Syme Thomson, I once more stumped up the landslide in the faint hope of being able to get through to them, but at the narrow ledge a tragic sight met my eyes. Red-flushed billows of smoke were swirling over the cliff-top and out of the murk figures scurried, scrambling over the edge in a frantic stream jostling, slipping and sliding down the improvised rope. This was the stampede that we had feared. Any hope of contacting our missing companions vanished and I stood aghast, wondering what would happen next. It came with horrifying rapidity. As the heat began to scorch the clothing of those still left on the terrace, now mostly Japanese from the villages, panic ensued and they started rushing the rope and surging over the cliff's lip anywhere they could, clutching at insecure handholds, slipping, recovering, slipping again, until they clung like barnacles on the brink of the last sheer drop to the beach. Under terrific strain, the tennis nets parted, leaving those on it suddenly without support, clawing desperately at any protruberance in the broken trail.

Tongues of flame now laced the smoke; and at that some of the figures silhouetted along the cliff-top began to hurl themselves over. At first I thought they were bundles of salvage being tossed over, but the horrible truth was revealed when one spread-eagled in mid-air and came shrieking down to sudden silence. Others followed, like rag dolls, helplessly pinwheeling down the cliff face and dislodging yet others until there was a motionless fringe at the foot of the cliff. There a number of fellow-Japanese had assembled, some of the younger men nipping in and carrying out the living, one by one. It was too gruesome and I retraced my steps downward, sick at heart.

Part way down I came upon an elderly Japanese man seated on a mound, head slumped forwards. I put my hand on his shoulder to offer help and found he was dead in that strange attitude of rest.

By now the reclaimed ground was being swept with stifling hot smoke from which we retreated outwards towards the water's edge. Everywhere were long fissures three or four feet wide, and the concrete blocks of the sea-wall had been tumbled into the water in confusion. A couple of thousand Japanese had succeeded in finding sanctuary on these few acres of reclaimed foreshore, most of them probably from Motomachi by way of the canal banks, and many of the wounded were being placed in a long bathing shelter made of rushes, which offered some protection from the heat overhead.

As in many ports in the Far East, one knew practically every other foreign resident, British, American or Continental; and now, in the ebb and flow of tattered figures, we began to meet acquaintances and friends, dishevelled and half stunned by what they had been through, and filled with grief for someone lost or fear for dear ones of whom they could get no news. Nearly all were cut and bruised, and some were more seriously injured. Rothwell Bowden (with whom I had done many a mountain climb in the past) was one of the first I ran into, helping his wife through the throng. They had been together in the Settlement but could not get up to their house at No. 45 Bluff. It was not until next day that they learned their children had been found with their Chinese amah in the near-by Bluff Gardens, unhurt, their house not having collapsed though badly shattered.

In the furious wind, fire and smoke was pouring over the cliff-top like the Niagara Falls; then half-way down it shot straight out in a flowing river close overhead, leaving us in a strange brown half-light. Retrieving one of the hospital sheets we had tossed from above, we managed to make a windbreak for the children who were almost overcome by heat and smoke and parched with thirst—as everyone was. It was a gift from heaven when someone discovered that a steel lighter lying a few feet off shore was a fresh-water barge. Quickly a gangway was contrived and water was hand-pumped to all who could bring a bottle, can or even a hat.

We were all exhausted by this time and we stretched out gratefully on the rough grass. Our three small boys had behaved like soldiers in every ordeal doing what they were told without fear or question; but now, with the rushing hot smoke overhead, David, only three years old, began suddenly to rock and moan inconsolably, in a sing-song chant. Realizing that this was a form of shock, as he seldom cried, Dorothy gathered him in her arms and gradually brought him round; whereupon, with a few sniffles, he fell asleep. Tony and Dick were also told to snatch a nap in the lee of the sheet but it did not come easily. Dick kept asking plaintively, 'But what will become of all the animals?' This was real heartbreak for him, because he loved them and was devoted to his toy creatures. Two years later in America, Dorothy found in a toy-shop the identical stuffed 'Puss-in-Boots' he had loved above all, and his joy in the belief that it had come through the earthquake safely was touching. Even Tony, who had stood up manfully to everything, leaned on one elbow and murmured wistfully, 'I wish it was all a dream!' 'Second the motion!' added the Commodore, fervently.

But the nightmare continued. Some poor devil, a Japanese who had fallen on the cliff and had been caught in the fork of a tree about fifty feet above the beach, was pinned there, shrieking horribly without pause. Inaccessible from below, the fire cut off help from above; and there he hung, yelling like a fog-horn, until dusk, when they were at last able to lower a rope from above and release him.

With all the world seemingly in a macabre state of flux, inaction was intolerable; and anxiety as to the fate of the rest of our party sent me searching through the hundreds of people standing or sitting in forlorn groups, while others still streamed from the mouth of the canal and the burning foreign Settlement beyond, a quarter of a mile from where we were pitched. Among them I came upon many friends, exhausted and unnerved. The J. R. McKenlays, without their baby. Mrs John R. Geary, with a broken ankle, had escaped from Motomachi with the help of her chauffeur and, seeing the flames above, was despairingly convinced that her children had been burned in their house next to Mrs Syme Thomson's. Mrs Edward Rogers, who also lived beside her and had escaped through the British Naval Hospital, could tell her nothing about the children, but it turned out that they had been saved by the amah in one of the adjacent big gardens. Resting on the ground was old Dentici, the French baker, with a badly injured hand; also C. K. Marshall Martin, the well-known collector of Japanese art, who had helped dig out the French Consul, Paul Dejardin, from the Consulate, only to see him succumb. There, too, were the daughters of Maurice Russell who had last seen him up above and feared the worst. Also D. E.

Yarnell of the Y.M.C.A. vainly exhorting distracted people to 'organize'. It was a haggard concourse, many almost unrecognizable with grime. And nowhere among them any trace of those I sought until I eventually got back to our encampment and there they were, safe and sound, among us again, lying prone on the grass to rest their weary limbs.

It seems that when Anderson and Bateman awoke to the fact that they were almost cut off, they raced back with such food as they had unearthed, only to discover that the mad rush for the rope had already begun, and there was no prospect of getting near it. Edward Coutts had vanished. Roused almost from unconsciousness by Rowbottom and a couple of sailors to a comprehension of the imminent danger, he had managed with their magnificent assistance to struggle to the clifftop and slide down the rope with his feet on Rowbottom's shoulders to safety below. But the Russian governess with the broken leg had had to be abandoned, and Anderson and Bateman carried her up a knoll beside the flagstaff a little beyond the Summerhouse, to which another rope had been made fast, leading directly into the gap of the landslide. Down this, in the general wild scramble, Bateman, his wife and Mrs Syme Thomson managed somehow to slide safely; but by the time Anderson had contrived a shallow trough under a bush for the governess, with an improvised fire-screen, where she again had to be left, fire had taken hold of the wreckage around the flagstaff and severed the rope, so that he, Honor and Patsy had to go down the precipitous scar of the landslide without any aid whatever. Frank went first in short stages, kicked a foothold and then caught Honor and Patsy as they slid to him. How they escaped falling they scarcely knew themselves, but they made it. As for the governess, she was rescued next morning, scorched but alive. Up by the flagstaff, Bateman and Anderson also saw old Maurice Russell, very portly with age, and had tried to per-

8. Looking across the Creek to the mouth of Camp Hill. The American Naval Hospital stood on the sky-line. In the very early days, this gully was the site of the French Garrison encampment. In later years the French Consulate stood just to the left, and the Consular residence on a spur behind it. The Japanese village filling the heart of this gully was completely wiped out.

Looking up the Creek from Camp Hill Bridge. Maida-bashi, the second bridge, is seen in the distance. Honmura and Motomachi, on the left of Maida-bashi, have disappeared.

Profile of Sengen-yama where The Hundred Steps had descended to Motomachi. They were completely carried down in the quake, together with some of the trees and the teahouses which had been on top, smothering the houses below. The single tree and trail on right mark where Motomachi had been.

9. Homes that escaped the fire. Residences of R. C. Bowden and W. Hayward and families at No. 45 Bluff.

A. P. Scott's house at No. 38 Bluff beside the 'Bluff Gardens'. This illustrates the term 'pancaked' that applied to so many houses.

All that was left of the author's residence at No. 68 Bluff. Slabs of turf on the bank left of centre show how the lawn slipped into the gully. Ruins in foreground are of L. Watson's house.

10. Bluff Road from Morris Mendelson's house, No. 103 Bluff. Behind the telegraph pole may be seen the ruins of Mrs Syme-Thomson's house, No. 108 Bluff. Also the ashes of A. E. Bateman's house opposite. The road to right leads down into Amanuma Village.

Escape route from the British Naval Hospital down the cliffs. (The camera having been pointed upwards reduces the height and steepness.)

11. Ruins of Grand Hotel in left foreground. Refugees on Bund at right. Smoke still rising from burning lighter at French Hatoba.

Walls of the Yokohama United Club at No. 5 Bund. Arches of the main hall at right.

Patched up pier and shell of the Harbour Office and Water Police Station. Seen over the ruins of the Hongkong & Shanghai Bank.

suade him to attempt the descent; but he was convinced it would be suicidal and, although alive to the hazard of staying on top, elected to take his chance, shook hands with them and walked away. His body was found there next day among others who had been overcome.

All that I have so far narrated had taken only three hours to happen, three crowded hours since the first shock at three minutes before noon. It was now mid-afternoon. There ensued a period of enforced idleness, since no move could be made until the fires subsided; so we put our heads together to decide what to do next. Beyond the milling throng one could see down the length of flaming Bund and across the harbour to Kanagawa, which was also burning. It was plain that Yokohama was being wiped out—not merely the Settlement, the residences on the Bluff and the Japanese city, but the surrounding countryside as well. There would be no roof under which to lay one's head for miles around; in fact, at that stage it was believed that all Japan was involved in the cataclysm. There would be no food nor drinking water, and there would be thousands of homeless to care for. Already it was being realized that our well-ordered lives had been dislocated beyond possibility of early resumption, if ever, and that we should have to plan accordingly.

As if to drive it home, our good friend Don Jose Caro, Spanish Minister to Japan, who had often dined with Dorothy and me, approached us at this moment with the information that he had passed our house in his flight, that it was burning fiercely and by now had undoubtedly gone. Singularly, the news left me unmoved. What did it matter that we had lost everything? That we were all alive and unhurt was more than our share of good luck.

There was great activity among the ships in harbour; some had already steamed out through the breakwater and dropped anchor a mile out in the bay. Others were following suit as fast as they got up steam, but navigation was tricky in the high offshore wind. So far none had sent lifeboats ashore; this was impossible in the teeth of that wind. We could see the *Empress of Australia* still tied up to the pier from which she was to have sailed at noon for Vancouver. In the yacht anchorage off the Y.A.R.C. boathouse, half-way down the burning Bund, could be discerned the Commodore's beloved *Daimyo*, gleaming palely through the smoke among the other yachts that were to have raced that afternoon; and seeing her so serene in the midst of chaos, our little party gratefully fell in with the Commodore's determination to find some way of getting us all aboard before nightfall.

The Grand Hotel and other buildings along the Bund were still vomiting red flames out over the water; and the old Pacific Mail coal sheds on the near side of the canal mouth had also caught fire from the French Consulate behind them, adding a new outburst to the spectacle of destruction. But by half-past four the Settlement was beginning to burn itself out, and Willy-Wally, Frank and I set off in that direction to see if our plans were yet feasible, leaving Bateman with the ladies and children.

Fugitives were still stumbling in by twos and threes, and suddenly one stopped short. 'Poole, is that you? I can't see you, but I know your voice!' I looked at a tragic figure in burnt pyjamas, his hair and eyebrows scorched off, his face blistered and his eyes red slits, and asked who in God's name it was. 'Don't you know me? I'm Starr. How are my eyes? I can't see.'

I took his arm and led him back to a place of safety near the water among the other helpless ones, while he told me he had been asleep in the Grand Hotel after a late night of cards and had been overwhelmed in the first shock, regaining conscious-

66

ness to find himself buried under fallen timbers. Presently a tiny aperture of light showed above, towards which he wormed persistently, finally emerging to discover that he was perched on top of the ruins of the hotel, already afire on all sides. His only way of escape was through the flames and he didn't know what had happened after that. I made him comfortable, assured him relief would be along soon, and that we ourselves would be coming back and would watch out for him. I never saw him again, but I heard he recovered after a long time in hospital.

Rejoining the Commodore and Anderson, we pushed on to the bridge over the mouth of the Canal, where we were glad to observe that the wind had shifted and was now blowing away from us straight down the Bund, so that the flames were no longer sweeping across it but raking the ruins parallel with the water. It thus looked possible to get along the Bund to the boathouse, whence the Commodore was determined to swim to the *Diamyo* if necessary and return with her dinghy, which could be seen alongside. Sending Anderson back to report what we were about to attempt, my father-in-law and I crossed the bridge to the crackling ruins of the Grand Hotel and headed down the Bund, but found the surface burned brick red by the intense heat of the fire and still too hot to tread on without burning one's shoes. However, we were able to creep along the ragged edge of the bunding until the water grew shallow enough to wade; then along a sliver of sand to what was left of the boathouse—a tumbled stone parapet, head-high. Vanished were the pavilion and all the boats and gear; and beyond was the equally bare curved camber of the French Hatoba.

The Settlement itself was a terrible sight. Beyond the tangle of fallen electric power lines and poles, snarling the Bund, stretched nothing but a vast bed of flaming embers, cross-hatched by shattered walls of brick or stone. All that remained of the five-storied Oriental Hotel, just across from the boathouse, was a cluster of smoking brick walls. We learned later that its French owner, L. Cotte, and half the guests were killed in its collapse.

Not a living creature was moving anywhere; this was a deserted world. Then, as we soberly took it all in, an uncanny thing happened. Over the parapet above our heads, silhouetted against the swirling smoke and crimson flames, slowly rose a swaying black figure, then another and another, like spectres of the damned emerging from a fiery hell. For a moment they were poised grotesquely against the lurid backdrop, then lurched silently down to where we stood. Harry Rankin, Capt. Rennie Tipple, W. G. Feast of Butterfields, H. E. Gripper of the Rising Sun (Shell Oil) and one or two others I cannot recall. Wet, black, bloodstained and numbed, they told us they had been among survivors of the first shock who fought through to the Bund, only to be driven off by fire into the shallow water between the United Club and the Boat Club, where ever since they had been standing shoulder deep, ducking their heads when the heat became unbearable. To them we were the first sign of life; and, cheered to learn that many others had escaped and had congregated on the reclaimed ground, they waded off along the Bund wall to link up. We noticed that Feast, whose head was crudely bandaged, was no longer with them, and looking back, we saw him wandering aimlessly out to the end of the camber of the French Hatoba, where he squatted by the water's edge, a picture of despair. We concluded he wanted to rest. Unhappily, he never was quite himself again; he had to be sent to England and died a year or two later.

Of course, there were some humourous episodes in the earthquake. During the first shock, the Oriental Palace Hotel on the Bund, opposite the French Hatoba, disintegrated, though some of its upright walls remained standing. On the third floor, a well-favoured young French woman, whose name it would be unchivalrous to reveal, was enjoying a bath. When the turmoil had ceased, she found herself still in the tub, suspended in the air by its plumbing but nothing below, only fragments of walls on either side. Her cries reached Stephen ('Tiff') Lucas, who had bolted out of the Standard Oil offices at No. 8 on the Bund; and being an active tennis player and in good trim, he dashed to the rescue. With goat-like agility, he scaled the walls by their ragged tops, a remarkable feat, and found himself beside the damsel in distress. But, the story goes, she was still slippery with soap, and try as he might, he could not get a grip. Resourcefully, he took off his jacket, buttoned it around her and lifted her bodily to his shoulder; then, balancing this alluring and clinging burden, he achieved an even more remarkable tight-rope descent of the brick walls. Most of those in the hotel had been killed, and since he was now her sole protector, he could not casually dump her and flee to safety, so he gallantly carried her into the water and there they bobbed about for hours, with only ripples to conceal her predicament, and 'Tiff' pink with embarrassment. Later, odd garments were contributed by others in the water, and 'Tiff' eventually escorted her to safety in the kindly dusk.

Not having witnessed the foregoing incident, I cannot say to what extent this story was embellished as it circulated aboard ship next day; but the rescue did occur and 'Tiff' was its much envied hero.

Now to resume where the Commodore and I were wondering how to get aboard the *Daimyo*. He was about to swim off to her when old Ichi the *sendo* (Sailor), who had been readying her for the afternoon race when the earthquake struck, suddenly appeared on deck. At the Commodore's well-known

69

hail, 'Honk, Honk!' he sped ashore in the dinghy, touchingly delighted to see his skipper alive. With Ichi trailing us in the boat, we two hoofed it back along the Bund, not too good an idea, as the treacherous wind veered off-shore again and we had to make a final dash through suffocating heat to a boat landing at the mouth of the Canal. Here we left Ichi and the dinghy, pushing on to our party, who received with joy and relief, the news that they were to embark at once. With the children on our shoulders, we all set off—that is, all but the Batemans and Mrs Syme Thomson, the ladies' injuries having become too painful for walking. So we arranged to make a second trip and bring the dinghy in among the rocks, close enough to carry them on board. Mr and Mrs King, who had rejoined our party, also remained behind, fearing we should be too many. At the landing we found and impressed a stray sampan, the reluctant *sendo* thawing at the promise of ten Yen. We were thus all able to embark at once, Dorothy and Miné with our three boys, Frank, Honor and Patsy Anderson, Mr and Mrs Campbell and myself. How wonderful it was to scramble on board the *Daimyo* and stow everybody away comfortably; and a stiff peg of whisky all round never tasted better.

Leaving Anderson in charge, the Commodore and I set off for shore again in the dinghy, allowing Ichi to take the sampan and try to find his family up the canal. I have never heard whether he succeeded.

The light was failing as we got back to the reclaimed land and we had some difficulty in locating those left behind as the foreign ships had been able at last to start their splendid rescue work and refugees were herding at a spot near the water's edge where lifeboats from the *Empress of Australia, Andre Lebon, Steel Navigator, Dongola, Benroich* and others had nosed in between the rocks as far as they could get, and stretcher parties were taking off the wounded first, then filling up with any who could pack themselves in. The collapse of the bunding everywhere made the work of embarkation difficult, as the lifeboats could not get in close enough; and seeing that this was where we could help, we brought around the dinghy and in the gathering dusk ferried many of the wounded through the half-submerged concrete blocks alongside the lifeboats, where strong arms lifted them aboard. Plucky Mrs John Geary with a broken ankle was one I had to carry; and if any man wants a tricky job, let him try standing in a bobbing dinghy with a healthy young matron in his arms.

It was almost dark when we rowed back to the *Diamyo* with Mrs Syme Thomson, Mrs King and Gladys Bateman; and I then made another trip to the boathouse slip to pick up King and Bateman who had walked around. Meanwhile, lifeboats continued their rescue work far into the night. Strangely, none of the Japanese cargo and passenger ships that had steamed out of harbour and lay only three-quarters of a mile off-shore participated in the rescue work that afternoon, nor lowered a lifeboat until next day. Whether the commanding officers were ashore and those on board unwilling to assume the responsibility of saddling themselves with refugees, or whether they simply did not rise to the emergency, is hard to say; but from the reclaimed land we could see their crews leaning on their elbows along the rail, their lifeboats idle in the davits. Naturally, the first thought of crews from the foreign ships was to rescue foreigners in distress; but they also took aboard as many injured Japanese and their families as could be accommodated. Of course, this still left hundreds to get through the night as best as they could and await relief from the authorities on the morrow. And with daybreak help did come to the limit of the means at hand.

Another cause of speculation was the puzzling absence during the first two or three perilous days of any Japanese warships from the powerful Naval Station at Yokosuka, only fifteen miles down the bay. Rumours spread that the entire arsenal had been so damaged in the quake that channels had been blocked and many of the vessels moored to docks close under steep cliffs put out of commission, some even sunk. That Yokosuka was in flames could be seen from Yokohama. When warships did show up early in the next week, they rendered willing and efficient help. It later transpired that the small town of Yokosuka was completely destroyed and 700 were killed.

Seeing the overcrowded state of the *Diamyo* and that another large yacht, the *Azuma,* belonging to George Moilliet, Vice-Commodore of the Yacht Club, was riding close at hand with no one aboard, we transferred to her Frank, Honor and Patsy Anderson and Dorothy and our three boys. Then, as it became evident that Mrs Syme Thomson and Gladys were in increasing pain and should have medical attention, the Commodore and I took both the yachts' dinghies and rowed the two sufferers and Bert Bateman over to the *Empress of Australia,* still tied up to the pier, where I helped them climb the gangway to the brilliant decks above, while the Commodore lay off with both dinghies. He could not tie up to the gangway as lifeboats were pulling in fast with full loads of refugees from ashore.

The decks were a stirring sight. In the blaze of electric lights and rich appointments, which struck one oddly after the red-lit desolation ashore, moved a motley throng of soiled, half-dazed refugees who, six hours earlier, had been the well-conditioned inhabitants of prosperous, old-fashioned Yokohama. They gave one the extraordinary impression of intangible people seen in a dream.

At every turn I ran into some old friend who in most cases had escaped from the Settlement along the pier. One of the first was my good friend from bachelor days, Charles Rice of the Hongkong & Shanghai Bank, who, after a joyful greeting, told me that Winnie and their children, Nancy and Ken, were down at Kamakura, fifteen miles away on the seashore, staying at the Kai-hin-in Hotel; and that he was determined to get down there on foot first thing in the morning. Meanwhile, he could only hope and pray that they had come through safely. He himself had escaped from the bank unhurt, though the building had broken up badly and some of the others were injured. R. C. Edwards' right hand had been crushed and most of it had to be amputated. Stinie Morrison was overwhelmed and his leg broken. They carried him to the garden gate to the Bund and left him while doing further rescue work; but he must have dragged himself away, for when they returned he was not to be found. Next day his body was discovered around Jardine's corner under a cart, where the fire had caught him. Mrs Purington, who was in the bank when the earthquake struck, saw her seven-year-old daughter and nurse killed by the falling granite portico where they were waiting for her; her husband and nine-year-old son were killed elsewhere.

So it continued. One's eyes grew moist at every handclasp. Until then I had believed that of all Yokohama's residents, only a few had survived; yet now familiar faces were popping up on all sides, dirty and worn but still full of fight. Most of them were seeking others and they crowded round me. 'Poole, did you see my wife anywhere?' 'Was my house down?' 'Did you hear anything of my children?' Among them was Nicoll of the Chartered Bank, with whom I had done several mountain trips, now stunned with grief; his wife and small daughter Helen had been killed in their car at the door of the bank. Only three days earlier I had been playing tennis with her. His little baby at home was saved by their Chinese amah and was brought in next day.

Just then Frank Shea of the American Trading Co., who was

aboard as a passenger to America, drew me into the smoking room and got me a good stiff drink, which I needed badly for meeting these old friends suddenly was curiously moving. I suppose I was a bit more shaken than I realized by all we had been through. But the Commodore would be waiting with the dinghies and I could not linger.

Outside was pitch dark and the shore was a continuous line of spouting flame from Honmoku around to Kanagawa, Tsurumi and Kawasaki. As we rowed back in the dinghies, like black-paper silhouettes over a mirror of reflections, the oil tanks out at Hiranuma were catching alight and filling the smoky heavens with new columns of billowing fire. The crackling roar of flames still assailed our ears and it was uncanny to be silently tugging at a pair of oars in the midst of it all.

It was only about half a mile to the yachts, where to our surprise we were cheerily greeted by Moilliet himself, whom we had imagined to be in Tokyo as usual and most likely cut off. It turned out that he had come down to see some friends off on the *Empress of Australia* and had only just been able to get to the *Azuma* bringing with him Norman Brockhurst, his wife and another man. Fortunately the *Azuma* had a commodious cabin with four berths and a settee, in which the women and children all disposed themselves comfortably for the night, while we men curled up on deck without much thought of sleep, or sat talking in low tones in the cockpit watching the last of Yokohama burn away to embers through the long night.

Bit by bit I learned from Moilliet what had happened alongside the *Empress of Australia*. If the earthquake had to come, it was a mercy that it struck when it did, for three minutes later the gangways would have been pulled away and the ship cast off. Had the two or three-score of residents who had come down to see friends off been in their own homes or offices, many would have perished beyond a doubt. As it was, most were on the pier, and a few just leaving the ship, when the earthquake threw them to their knees. The pier started to break up, the long sheds in its centre subsiding bodily with their heavy cement floors into the water. People clung to the open planking of the pier, but some were flung into the water or were carried down into the pit together with a few automobiles that were parked on the pier and skidded down the sloping timbers. Some were killed or drowned, among them Mrs H. W. Taylor, whose husband was in the C.P.R. and I believe on board despatching the ship. Sailors from the *Empress* and *Andre Lebon* immediately rushed to the rescue and everyone was taken on board, since the pier had become all but impassible and the Settlement could be seen to be in ruins. Captain Robinson, standing on the *Empress'* bridge when the earthquake struck, reports that he felt the ship shudder as if she had run aground; and in the same instant saw all the buildings on the Bund, only a few hundred yards away, dissolving as if made of sand and the Bund itself heaving in waves six to eight feet high! From the crashing ruins there instantly rose a dense cloud of yellow dust to a height of three or four hundred feet, blotting out all further view of the levelled city; and it was not until the dust settled again that he could see the terrible havoc wrought. From that moment he decided to remain at the pier and become a relief ship. Strangely enough, only two weeks earlier in Hongkong, he had similarly made the ship a haven to hundreds of Chinese victims in the midst of a ferocious typhoon; and the name of the *Empress of Australia* will long be remembered by those she rescued in both ports.

I shall never forget that night. Lying on the hard cabin roof, with a stiff fold of damp canvas pulled over me, I had only to open my eyes to see the fire burning relentlessly hour after hour. Fuel oil from tanks or lighters was gradually spreading throughout the quays and cambers of the Customs compound; and as each fresh drum or lighter exploded with a deep 'Hrrrump!', a volcanic rush of gas and fire mounted upwards, unfolding like a succession of mushrooms, each exploding anew, until the last billow lost itself in the dense canopy of driving smoke. All the fireworks I have ever seen were insignificant compared with this stupendous and furious spectacle.

None of us succeeded in snatching much sleep. Dorothy, ever watchful of the children, crept up now and then to sit shoulder to shoulder with me against the mast, watching the continuing drama ashore. In the enveloping summer night, the relentless roar of flames sounded like heavy surf, with frequent crashes of thunder. We seemed to be in the centre of a huge stage, illuminated by pulsing, crimson footlights. Throughout the incredible day, Dorothy had met danger with unwavering courage and resourcefulness. Now, safe in our temporary haven, we could for the first time reflect sadly on what it all meant—the shattering of our happy existence in a place we had both known and loved since childhood.

In the darkness of the night, we could see a thin rim of fire all around Tokyo Bay, meaning that fishing villages and small towns were all sharing the same fate; the glare above Yokosuka, where the jaws of the bay come close together, showed that the Naval arsenal was also going up. Northwards over the water there rose on the horizon a billowy, pink cloud like cumuli at sunset, so distant as to seem unchanging and motionless, yet each time one looked it had taken a different shape. This was Tokyo burning, and by the cloud's titantic proportions we knew the whole city must be in flames, as indeed most of it was.

As we sat there talking quietly, the enormity of the disaster sank in. This was no incident of a day but a cataclysm. Yokohama was gone and the old life with it—business, homes, friends and old associations. We who had survived would have to be evacuated and begin life afresh; where and how there was no telling until we were once more in contact with the outer world and knew how far the calamity extended.

To court sleep was almost impossible. Fears haunted us for relatives and friends behind that wall of flame. What had befallen them? Calla Campbell's widowed sister Mabel Fraser was to have taken the train at a quarter past twelve for Miyano-shita, so would have left the Antoinette Apartments beside the Grand Hotel before noon. Where had the earthquake caught her? She must have been in the very heart of the city.

And what of my father Otis A. Poole, the sturdy seventy-six year old doyen of 'Tea-men' down in Shidzuoka by the tea-fields at the foot of Fuji-yama. I pictured him in his Samurai's residence beside the ancient moated castle and wondered if its white towers still guarded the medieval city or looked down upon another sea of ruins. For my sister Eleanor Maitland and her youngest son Donald, with faithful Emily, over from Shanghai to summer at Karuizawa in the mountains of central Japan, I had less anxiety, though their proximity to the active volcano Asama-yama injected an unpredictable hazard.

From one close friend to another, our thoughts roamed apprehensively, while the sparks soared in eddies overhead.

Throughout the night, lifeboats plied back and forth between the ships and shore, and it was nearly daybreak before the last boat-load of survivors from the reclaimed ground came by, with the Reverend Eustace Strong in the stern, head bandaged and worn out. He had done splendid work combining the shore for isolated groups and shepherding them to the boats; and in the following days he proved tireless.

Finally, with the sound of oars in our ears, we fell asleep.

❧ III ❦

SUNDAY

September 2

At the first hint of dawn, I rowed over from the *Azuma* to the *Daimyo* to confer with the Commodore. It was plain we could not see this through on the yachts. For one thing, we hadn't enough food; for another, conditions around us might become hazardous. The conclusion was inevitable: we should have to face facts and throw ourselves on the hospitality of the ships in port until we could be evacuated to some other port in Japan or the Far East or even America.

In the pearly light before sunrise, we therefore collected our party and rowed over to the *Empress of Australia,* where willing hands helped us aboard and escorted us to the dining saloon, at one end of which a simple but wholesome breakfast was being doled out into mugs and soup-plates to a line of famished refugees. Then presently we settled the women and children wherever we could find an unoccupied corner. Later, they were kindly given cabins.

Weary stragglers who had lived through the conflagration in the park, or submerged in canals surrounding the settlement, were now climbing aboard, to be warmly welcomed by friends milling around the decks. Many had had terrible experiences or narrow escapes, and those with wounds were hurried off to the doctor's quarters for treatment. None of all those we questioned had seen or heard anything of Mabel Fraser and our fears rose. At every moment we heard of other old friends who had lost their lives.

The worst news concerned the Yokohama United Club. Though a massive building of brick and granite, four or five stories high, on the Bund facing the Bay, it had gone down almost instantly. G. Homewood of Shell Oil, standing on the colonnaded front verandah, leapt over the rail to the Bund eight feet below and got clear, but Eugene Fox, Secretary of the Foreign Board of Trade, standing near him, was flung back against the building and held there helpless while bricks and stone cascaded down before his face like a waterfall, the backwash piling up around his legs till he was buried to the waist and only just able to extricate himself, painfully hurt. Of nearly a score of members who were in the long bar—the first cheery arrivals of the hundred or more who would have gathered in another fifteen minutes—not one escaped alive. They had stood rivetted for a few seconds in incredulity, then turned to bolt for the front hall and verandah, when the whole Club fell in and crushed them. This is known because the wall behind the bar, with three tall, arched windows, fell inwards, one of the open windows dropping like a horse's collar over 'Jimmy', the popular Japanese head bar-boy, leaving him standing in the open air miraculously untouched. He knew and remembered who had been at the bar and what happened to them before his eyes. Two weeks later, Jimmy was installed behind the Kobe Club Bar, still with a single lock of black hair plastered across his bald pate from ear to ear, and a welcoming grin and extra-generous dollop of whisky for his old Yokohama 'danna-san'.

In the week following, Pat Lefroy's body was found on the verandah steps. 'Jock' Watson, 'Robbie' Robertson and 'Paddy' Ackland were found in the hallway just outside the Secretary's office. Not unearthed for some time were W. B.

Mason, one of Yokohama's earliest residents; Robert Gill, G. S. Niven, Cullinan, Brigel and several others whom R. M. Clarke, the Secretary, walking through the Club a moment earlier, remembered having seen. Of sixty-five Japanese 'boys', only twenty-seven escaped alive. On that Sunday morning all was vague and harrowing, and the blotting out of these popular men was felt poignantly by everyone.

With these sad tidings came word that Yokohama's beloved doctor, 'Judy' Wheeler (so called for his resemblance to 'Punch') was missing and believed lost. In the 1860s, when a doctor on a British warship, he was attached to Sir Harry Parke's retinue in the British Legation in Tokyo. He elected to remain on in Yokohama and became an institution in the community, bringing over 1,000 babies into our small world, including my wife and our three children. In the old days he used to end his morning rounds by enjoying a sherry at noon with his crony John W. Hall, the auctioneer on Main Street in the Settlement; and when the latter died in the fullness of time, Tom Abbey, who inherited the business, carried on the tradition. Thus it was that, just before noon that Saturday, they were waiting in the Auction Rooms at 61 Main Street, glasses in hand, for the noonday gun, when the earthquake struck. Havoc followed, but both were spared and tried to escape on foot to Camp Hill bridge and thence the short distance to the doctor's house at 97 Bluff on a spur above the bridge; but the old doctor with advancing years had become very heavy and unsteady on his feet, and, unable to move faster than a snail's pace, they were overtaken by fire in Macy's open tea-packing lot near the P. & O. and perished there with six or seven others from suffocation and subsequent fire. They had been seen together when setting out, Tom Abbey helping his dear old friend over the piled debris in Main Street; and still together, their charred bodies were found.

12. The canals and docks were filled with corpses of those unable to escape the holocaust.

Some of the bodies of those who took refuge in the Yokohama Specie Bank Building.

Fire breaking out in the Japanese city and customs compound as seen from the *Empress of Australia* twenty minutes after the earthquake.

13. Scene of horror. 32,000 people hemmed in and burned in the grounds of the Army Clothing Department in Tokyo.

What happened to country roads.

Temporary graves of foreigners in British Consular grounds. Shells of buildings in the background stood near the upper end of Main Street.

Sunday

Among the weaving figures on the *Empress of Australia's* deck, I suddenly saw my lifelong friend, Morris Mendelson approaching, a deep cut on his forehead and both arms bandaged almost to the shoulders; but what was more distressing, he seemed partly stupefied and incoherent, asking over and over if I had seen his young wife Madeleine. I described how I had been twice to his house with Parkhouse and Frank Anderson and was certain she had not been there, but that we had since heard no word of her. He had been through a ghastly experience. Seated at his desk in Berrick Bros. office, only half a block from the corner of the park, a wall-cabinet directly behind his chair had fallen forwards, imprisoning him face-down on his desk with arms flung forwards, in which position the building collapsed over him, a beam pinning down both arms as in a vice. His loyal Japanese, who had escaped from their portion of the building, started digging down to him, but it was slow work and fire drew closer every minute. Finally, they freed his body but could not lift the beam. He smelled the smoke growing hotter and hotter and at last his men had to cease their encouragements and convey the terrible news that the fire was upon them and it was now or never for a final effort. They would all get their hands on one end of the beam and give a gigantic heave, and when they shouted, he was to pull his arms free, whatever the cost. They tugged, signalled and Mendy gritted his teeth and yanked out his arms, lacerating the flesh from elbow to wrist, but getting free.

With not a moment to spare, they all leapt clear of the fire and made their way to the Park, where, with many thousands of Japanese and a sprinkling of foreigners, they rode through the encircling holocaust, moving from one side to the other with each change of wind and keeping wet by dipping in a pond that had, mysteriously formed in the centre of the park. The wind, by this time almost a typhoon, bombarded them with burning shingles, futon (quilts), and they had to watch each other's clothing to see that it did not catch fire. Sometime during the night, when the flames had subsided, Mendy and some other foreigners made their way to the *Empress of Australia.* Now, after receiving first aid, he was desperately seeking word that Madeleine had reached some other ship. Poor Mendy, he had yet to learn that she and the chauffeur had been killed on the steps of Brett's Pharmacy on Main Street. She was a lovely girl and they had been married for less than a year.

There being nothing we could do on board, the Commodore and I were itching to get ashore and see what was left of our business premises and possibly contact some of our staffs. But our chief aim was to find some trace of Mabel Fraser. However, the officers of the *Empress of Australia* were reluctant to allow anyone ashore, as reports were filtering through of looting and bloodshed and obviously things on land were chaotic. But about 9 a.m. the Commodore and I, together with Charles Rice, snatched a lift in a passing lifeboat which dropped us off at the *Daimyo,* and after questing about in her dinghy, we managed to induce a fishing smack with a gas engine to convey Charles down the Bay to Tomioka or Nagahama, a six-mile leg of the journey to Kamakura. Just then we were hailed by another lifeboat which rested on her oars alongside and in the exchange of news gave us the cheering word that Mabel Fraser was safe aboard the P. & O. *Dongola.* Also that the Campbell residence and Charles Rice's, as well as four or five others in the same fold of the Bluff, had escaped the fire and were still standing. Meanwhile Rice had discovered that several in the lifeboat were also bound for Kamakura and glad of a lift down the bay with him. So there and then we effected a general post, Mackenzie, Loftus, Stott, Fitzgerald and August Manley joining Rice in the fishing boat, while the Commodore and I, with the others in the lifeboat, landed at the Grand Hotel corner of the Bund.

I might here interpolate that the Kamakura-bound party, aided by August Manley's fluent Japanese in some trying circumstances, got safely through the torn-up countryside and found their families camped outdoors in the gardens of the hotel or of their private houses. A few were injured but there were no reports of fatalities; whereas the Japanese in the shore villages suffered a heavy loss of life. A tidal wave had swept up every inlet, adding to the toll; but it had not been large enough to surmount the high dunes on which the Hotel and Summer houses were built. Charles found Winnie, Ken, Nancy and Miss Duncan unscathed except for a few cuts and bruises incurred when escaping from the falling hotel in which a number of Japanese servants were unfortunately killed. Before nightfall, the men had managed to erect a tent of sorts for all the women and children, made feeding arrangements and organized a patrol. In Dzushi, beyond the next headland, the tidal wave had raced up the beach and over the gardens of the foreign residences strung along the shore. Fanny Warming, her eight-year-old daughter Karin and their amah, in the back garden, were submerged to their waists and nearly sucked away in the backwash. The house had been flooded and they spent the next three days in a hut on the hilltops, as did others around them, and were found there next morning by Sophus Warming after an exhausting night's trudge from Yokohama, fourteen miles away.

On Thursday morning, five days after the earthquake, an American destroyer, *Smith Thompson*, steamed around from Yokohama and picked up all foreigners at Kamakura, Dzushi and Hayama, and before midnight had them all safely on board the *Empress of Australia,* still lying off Yokohama in Tokyo Bay.

To resume my narrative at the point where the Commodore and I had landed near the Grand Hotel. From the corner of the Bund, we could see at close range that there was no hope of entering the Settlement's streets. Our offices were still unapproachable because of the heat of the ruins. A lighter loaded with paint or some inflammable cargo had grounded half way along the Bund wall beyond the Boathouse, and in the onshore wind was belching a thick curtain of black smoke across the Bund, effectively closing that route. So we decided to head for the Bluff and see what was left of our homes.

Passing through knots of waiting refugees, mostly distressed Chinese, we trudged up the Camp Hill gully, now a hollow bowl of ashes, past one or two roadside corpses, past the abandoned auto, past Morris Mendelson's amah, sitting beside the body of her husband who lay where his injuries had overcome him, to the top of the hill between the British and American Naval Hospitals, and onto the Bluff Road. But where was the Bluff? Instead of the old familiar scene, or even the tossed debris of yesterday, there rolled today a wide stretch of bare hills and dales, devoid of anything but graceful clumps of trees scorched to Autumn nakedness. Patches of still green lawns glimmered here and there on the blackened slopes. But not a house in sight! Only patches of rubble as insignificant as the leavings of a thoughtless picnic party. For all trace of habitation, the hills and valleys were as featureless as the farmers' fields of seventy years ago, before the foreigner came to Japan. And, grotesque as it may sound, my first impression was, 'Except for the desolation, never has the Bluff been so beautiful!' So much for a sudden glimpse into a virginal past.

But then came the reaction: the pitiful cemetery with its

chaos of overturned headstones, and barren blanks where houses, familiar for years, had stood as landmarks of affectionate memory. The empty garden of 89 Bluff, where I had lived for nearly thirty years since childhood. And then the jagged, bare corner of No. 68, where Dorothy and I had come to start life after our honeymoon, and in which all our children had been born. Just the stumps of chimneys and folds of fissured lawn slipping down into the valley. On the grass, beside a small rockery, lay one of the old iron carronades from a Dutch barque wrecked in the the 1860s, its wooden carriage burnt away beneath it. Here, the twisted, rusty frame of Tony's tricycle. On top of the rubble, the Colt automatic that always lay on the bed table by my pillow, fused and reddened by intense heat. I still have it as a memento.

We poked around among the ruins, unearthing blobs of melted silver and glass, all that was left of our lovely wedding presents. And in one spot, where a Korean chest had stood in our drawing-room, filled with twenty-two albums of photographs illustrating thirty years of life in old Japan, just a neat pile of perfectly foliated ashes. For safety's sake, I had kept my negatives out in the stables, but they too had been burnt beyond redemption, as were all my records of countless trips and explorations up country in the mountains of Central Japan and among the disappearing Ainu of Yezo, the Northern Island. I may be forgiven a momentary lump in my throat as I gazed at the ashes of our first happy home, but a few breezy nautical observations by the sympathetic Commodore promptly restored my spirits. So, after a last look, we went on towards his place.

Beyond our house, the Bluff presented a spectacle of dreary desolation, every spur swept bare.

Near the top of Hegt's Hill (Daikan-zaka) we met a grand old-timer N. F. Smith, ageing father of my lifelong friend and best man K. van R. Smith, supported by two Japanese servants and headed for the water-front. As we halted to greet him, he seemed oddly dazed and to my enquiries after Van and his young wife Helen and baby Maureen, who all lived with 'N.F.' at No. 1 Bluff, he repeated dreamily 'Oh Van? Why, Van's all right. Yes indeed, Van's all right. And Helen? Yes, Helen's all right.' That was all we could elicit. Obviously, he was bemused from shock. But his familiar old house-boy told me quietly that all was far from well. It transpired that Helen had been sewing outdoors in the shade of the house, with little Maureen asleep beside her in a baby-carriage, and at the first shock, tiles began dropping on them from the roof. Before Helen could whisk the baby-carriage to safety, Maureen had been struck on the head and Helen herself cut and bruised.

Van had been in his office in the Standard Oil on the Bund when the quake came, and survived by standing under a doorway while the collapsing floors engulfed several of the Japanese staff in an adjoining room. Extricating himself with difficulty, he made for the Bund over the debris of the Oriental Palace Hotel and sped along to the Creek, crossed the fallen span of Camp Hill Bridge and up the hill before fire blocked it. Dashing along the Bluff, he spotted Dorothy and the boys with Honor and Patsy Anderson in the garden of No. 89 (I had not yet got there), and nipped in for a moment to see if they were all right; then pressed on to No. 1 Bluff, where the sad situation awaited him. His father had fortunately come through safely. There, too, fires were approaching from the village below, so they moved across the side street to the large gardens of 225 Bluff, Van going back to drag out enough bedding to see them through the night. Others joined them as things grew worse, including old Captain Carst and Torleif Jensen. When daylight came, they decided to make for some ship in harbour. Van and Helen, with the baby cradled on a board, went on ahead, leaving their reluctant father to be escorted by their trusted servants. Van and Helen succeeded in reaching the *Empress of Australia* at the pier, where Maureen was put under the doctor's care, 'N.F.' rejoining them later. I believe he was picked up by a lifeboat from the Bund.

Looking down Daikan-zaka there was nothing to be seen but the ashes of Motomachi, the jagged furrow of the Creek and the idly smouldering ruins of the Settlement beyond, misty with drifting smoke. Around us blackened trees burnt bare of leaves, and only the foundations of homes so familiar to us through the years.

A little farther along the Bluff road we met Hilare Tait, distracted at the non-appearance of her husband Arthur Tait, Manager of the Chartered Bank of India. 'I know he's dead, Chester, I know he's dead!' she gasped. 'If he weren't, he would have got to me somehow by now!' Poor girl, we tried to reassure her, but we felt in our bones that she was right. It turned out that Tait had been instantly killed in his rikisha on the way to the Club, overwhelmed by the collapsing brick walls of the Singer Sewing Machine Co. at No. 23. Either George Blundell or Cyprian Stanton, while clasping a telegraph pole at the next corner, saw Tait and his rikisha-man engulfed before his eyes, but it was not until the blocked roads could be systematically cleared many days later than his story was verified. Hilare herself had been sewing with her young daughter Jean and both were caught under the house as it fell, but were dug out by the faithful servants, hurt but not seriously injured, though Hilare was painfully lame. Several weeks later, a four-inch piece of knitting needle was discovered and removed from her thigh, a hand's breadth from where it had been driven in.

Beyond the Bluff Police Station, all that was left of the Union Church was spilled over the roadside in great masses of concrete, whereas Capt. Owston's ancient wooden house next door, built way back in the 1870s at No. 48 Bluff—(which I remember best as Mrs Smedley's School for Young Ladies back in the 1890s) stood intact though twisted, already a prey to looters. Beyond again, on the left-hand side, Roth Bowden's and W. Hayward's houses also stood, unburnt and still upright, but with their backs torn off. Then, around the turn to the Bluff Gardens, the badly fissured Catholic Church just breaking into flame.

A short distance down the Bluff Gardens lane, we stepped through a garden gate into the Campbell's drive and discovered the house just as they had left it the previous day, squatting on the ground like a crumpled paper bag, with a big slash down the middle. Through the windows the furniture could be seen in complete disarray. Out in the garden their servants had already improvised rude shelters to live in, their own quarters being wrecked beyond use. In fact, no one wanted to be under a roof as long as earthquake shocks continued every hour or so. There were twenty-two in the first twenty-four hours, unpleasantly pregnant with ugly potentialities, though they gradually diminished in strength. After a chat, we climbed into the living room through a window, finding the hardwood floors in waves like an oily swell at sea, and everything in dusty disorder, paper torn off the walls and bric-a-brac smashed. Looters had already been in, paying no heed to the servants who immediately afterwards took out and hid away in their shelters anything of value they could save, but were in great fear that the next looters would murder them to get at it. One or two precious keepsakes the Commodore decided to take with us, but the rest he left with the servants, telling them he would not be going away but would keep in touch with them until relief could be organized. So, without lingering, we started back to the main Bluff road.

There we ran right into Jeffery, whom I had last seen in

our office the previous day, together with Gripper and three or four others making their way afoot from the Negishi Golf Course area to the harbour. Gripper must have got around to Negishi from the reclaimed ground by way of Juniten and Honmoku, after we had seen him at the Boathouse the previous afternoon. Quite a number of residents from the Bluff had found safety in the Negishi hills instead of trying to get to the Bay, but the police were now warning them to make for the ships, as the Negishi jail had crumbled and gangs of escaped prisoners were already plundering and assaulting people.

It was good to see old 'Jumbles' (Jeffery) still in the ring, and as we walked along together I learned what had happened to the rest of our office staff after Bateman, Anderson and I had providentially dashed for the Bluff. It had only taken a jiffy to get the books and records into the vault, but the thick walls having been jarred askew, it proved impossible to close the vault door by several inches. There was nothing they could do about it, so they had to leave it that way; and all the contents were destroyed in the fire. Jeffery himself first tried to reach the Y.U. Club by way of Main Street and the Hongkong Bank, but foiled by impassable debris, he headed back and attempted to get through Silk Street and out by way of the International Bank. Silk Street was bad enough but the Bank and American Express Co. had crashed towards each other blocking that exit fifteen feet deep. Rose, Manager of the Bank, was a mass of blood with a cut head and broken arm; and two of its newer young Americans whom I did not know were killed outright. O. T. Gillon was there too, sole survivor of his office across the street. And F. W. Hill had succeeded in escaping from his rooms on the second floor of the American Express building by the simple process of walking down the heap of tangled timbers.

Seeing there was no exit that way and fire having broken

out in the basement of the Bank, Jeffery and the others beat
a retreat up Silk Street to our office, to be confronted by Siber
Hegner's office diagonally opposite ours already in flames.
Here he found James A. Thomson, John Barnett, Gordon Bell
and George Colton with our two stenographers Lucy Fox and
Miss Martin on their hands, the latter hysterical and virtually
having to be carried. All now concentrated on scrambling
through the last block of Silk Street to Satsuma-cho, the wide
avenue on which the Chartered Bank stood, and down it to the
Park. The Chartered Bank had fallen with tragic results, as
had also the Deutsche-Asiatische Bank on the corner beyond,
where Jeffery paused to aid an attempt to rescue P. Sandberg,
the popular manager, who was trapped. But they failed and he
perished.

Meanwhile the others had struggled on to the Park, crossing
many fissures that split the roads and wading though water
that filled every subsidence, gaining the wide open space in
the centre that in earlier days had been the peerless foreign
cricket and athletic field until the Japanese Government dis-
possessed them. Whether Jeffery found and rejoined them I
have forgotten, but in any case they all flopped down there
amongst thousands of Japanese who had fled to the same
haven. In an incredibly short space of time they were surrounded by flames on all sides and, in the near hurricane wind, had
continually to dodge burning fragments of buildings that
whirled and bounced around them. The worst hazard, however, was from spinning sheets of corrugated iron sailing
through the air. A Japanese woman in the crowd was decapi-
tated as if by a sword. When the flames got too hot, they dipped
themselves in a pond that had come into being; and so rode
through hours of strain until the city was burned out and they
could creep away, either late that evening or when daylight
came.

Jeffery worked along the canals and up Jizo-zaka to the west end of the Bluff, but the road thence through Aizawa village to the Race Course was still curtained by fire, forcing him to make a long detour via the new Y.C. & A.C. grounds around to the back of the Negishi hills and so the house in which he ran a mess with Gripper and Robbie Robertson of Shell, who was killed in the Club. The others in the park made for the harbour and succeeded in reaching various ships. Later in the day I came upon John Barnett and Jim Thomson quietly leaning over the rail of the *Empress of Australia* as if nothing had happened. They planned to await the arrival of one of our Swedish Asiatic ships then due from a southern port, direct the skipper to discharge her cargo at Kobe, and proceed in her to that port. George Colton, Gordon Bell, Lucy Fox and Mary Martin I did not glimpse again until we all reassembled days later in our Kobe office.

But to return to our meeting with Jeffery on the Bluff. By that time a stream of people was flowing along the Bluff Road towards Camp Hill and the waterfront, urged on by police and ugly rumours of an uprising of Koreans. True, there were many Korean convicts in Negishi jail; also hundreds of Korean labourers in the industrial strip between Yokohama and Tokyo, all malcontents; but these alarming tales of an uprising were a figment of panicky imagination. What unhappily seems true is that the police, fired by these tales, dealt summarily with any Koreans acting at all suspiciously, as well as all looters, Korean or Japanese, caught in the act, stringing some of them up to telegraph poles and shooting or executing others. I saw several knots of police with trussed-up prisoners, but none of the bloodshed. Others did, however, and powerless to intervene, said it was pretty awful. I suppose it was a type of martial law; certainly it put a quick end to disorders.

Among the fugitives on the Bluff, I came upon Henry Arias

of the Standard Commerical Tobacco Co. of New York, with whom I had done a large business during the war, accompanied by his wife and three-year-old daughter, all in tears. They had been living in the Fairmont at the edge of the cliff overlooking the city; the wooden structure had collapsed and caught fire, and although the three of them had got out into the garden, their baby boy was consumed in the flames before their eyes.

It was nearly noon when the Commodore and I reached the bund and decided to go first to the *Daimyo* before rejoining the family on the *Empress of Australia*. Ichi had been aboard again and the sails were furled and everything stowed away for a period of suspense. We were snatching a drink and a bite of lunch when, to our dismay, a new and unexpected danger suddenly assailed us. The burning lighter beyond the French Hatoba had collapsed, but up to now the wind had driven its flames over the Bund and held close to shore a scum of black fuel oil that had been flowing all night into the harbour from burst tanks on the low flats beyond the Custom House. Now, without warning, the wind swung right around to offshore, and in a moment the surface of the harbour was alight for a couple of hundred yards down the Bund, and deep red flames were advancing outward. They first swept down to the foot of the pier, then outwards, consuming the pier as they advanced. At the same time, tentacles from the lighter were reaching out towards us, too, and the sizzling roar as each puff of wind flung the flames forward another fifty yards, like groping hands, made one's heart thump. In a moment all was consternation aboard the ships at the pier and in harbour. We, too, were in imminent danger, so, without waiting to break out sail, we cast off our moorings and under bare poles ran before the wind towards the harbour entrance, only just ahead of the flames, until we reached clear green water and safety. Here, inside the mouth of the breakwater, the *Azuma,* a lighter, a small float-

ing crane and a sampan had just tied up to a handy buoy and
threw us a line.

Meanwhile, we had been helpless spectators of a frightening drama of suspense half a mile away. The *Empress of
Australia*, berthed at the pier bow shorewards, and the *Steel
Navigator* close behind, bow against the *Empress'* stern, were
right in the path of the oil flames advancing down the pier, as
was the *Andre Lebon* on the other side. The latter, helpless
with dismantled main engines, had cast off and was being
warped slowly by her winches to a buoy in harbour to which
young Tom Laffin in a small launch had daringly run a haw-
ser, dodging between patches of burning oil closing in on all
sides. Tensely, we waited for the *Empress* and *Navigator* also
to pull away, but although one could see commotion on both
ships, unaccountably neither stirred. Suddenly, a knot of
sailors on the *Navigators* bow began swinging what looked like
axes at her anchor chain. We did not know then that they were
knocking out a shackle of her port anchor cable which had
fouled one of the *Empress*'s twin propellers, so that neither
ship could get away from the other until the cable was severed.
What followed we could only watch with bewilderment.

Every moment the fire was pushing nearer and nearer the
motionless ships, and we could see the crew of the *Empress*
running out all her fire hoses and playing them on the ship's
sides and the adjoining pier. Still she did not move, and we
prayed in mounting fear for our dear ones and the thousands
of refugees in peril. The fire was now racing along the pier
and knots of Japanese fugitives, cut off and driven to its end,
shrieked for help. Little launches darted bravely in, taking
off a group here, another there. Tom Laffin, son of Tim
Laffin the ships-chandler, did courageous work here and saved
many lives at the risk of his own.

The advancing fire was now belching such volumes of black

93

smoke that the wild scene was shrouded in semi-darkness. And still the *Empress* did not move. It was only a question of moments now. Then at last, with a cheer, we saw her begin to pull away from the *Navigator* and creep slowly ahead. Her bow, blown away from the pier, was now pointed at the middle of the Bund and to our great relief we could see that she was now skirting the flames that were perilously close to her starboard side.

As the water shoaled rapidly towards the shore, we expected to see her stop and turn, but to our amazement she kept right on. Could her skipper be mad? She would be aground any moment now. Straight ahead, a group of lighters and a launch lay at a mooring in a corner of clear water on which the burning oil was fast bearing down. Could he be going to their rescue? What a risk! But to our horror the ship cut right through them, with a crashing of timbers and clamour of shouting as their crews were pitched into the water. Instantly lifebuoys at the end of ropes were flung from the ship's deck to the struggling figures in the water, and all who could grab one were pulled aboard, leaving the rest to the busy little launches to pick up. It looked like senseless massacre, and by that time we were convinced the Captain had 'gone off his rocker'. A minute later, with unnatural suddenness, she came to a stop. Was she aground? No; her bow began to swing slowly around in the strong wind. Gradually she was broadside on, then bow on, and we realized she was now coming dead at us. Surely, we thought, she would veer to one side or the other. But not a bit of it; on she came, right for us.

The Chief Officer, standing in her bow, raised a megaphone to his mouth and howled, 'Cast off that buoy!' The Commodore shouted back, 'Be damned if we will! Go around us!' Up went the megaphone again, 'Cast off, I say, or we'll run you down! We can't help ourselves!' Having seen the fate of the lighters, we needed no further spurring and instantly cast off, Moilliet springing to his gas engine in the hope of towing us all aside. 'Put-put' was all the response. 'Put-put-put' came at the next try. The *Empress* was looming bigger every second. 'Put-put-put-put-POP-POP-POP-POP . . .' and we inched away just as the huge bulk of the ship swept by, washing us aside like corks. 'Sorreeee', came a call from above and then we saw a long line of refugees' heads leaning over the rail and we exchanged waves. No sign of our own family, who did not know how close we momentarily were.

As the *Empress* continued on her course, we could see that in heading straight for us, she had also been heading straight for the harbour entrance, though at a dangerous angle, and were relieved, therefore, to see her slither through and out into the bay where she dropped anchor about a mile out.

It was only later in the day that we learned all the facts. We had witnessed a superb and now historic feat of seamanship by Captain Robinson. Though cutting the anchor chain of the *Steel Navigator* had separated the two ships, the chain still remained snarled in her propeller, so that, as she moved forward under the other engine and free propeller, she towed the anchor behind her like a tin can tied to a dog's tail, which made it impossible for her to answer her helm. She could go only straight ahead, unable to turn, a frightful predicament for a ship facing oncoming fire. But Captain Robinson's resourcefulness was equal to the emergency. Ignoring the proximity of the fire, he steamed straight for the Bund until he could see over his port quarter a clear path down-wind to the harbour entrance; then he let the wind swing his vessel around on the pivot of the trailing anchor until he was pointed at the entrance with the anchor strung out behind him, whereupon he steamed ahead once more, powerless to veer off the straight course. He could not avoid ramming us had we not evaded him; and with

5,000 souls on board, would not have hesitated to do so. Far from thinking him mad after that, our hats were off to him for a remarkable feat.

We were now among the yachts near the breakwater entrance. The harbour was filling with burning oil, so we concluded we had better get out into the bay too, as all the ships were doing. Moilliet kindly offered to tow us wherever we desired. He, too, had been ashore in the morning and had brought back to the *Azuma* his friend, Dr. Paravicini, with a broken ankle, and propped him up comfortably in the cabin. As there were ample stores on board, Moilliet planned to sail down the bay and around Uraga and Misaki to the little landlocked harbour of Aburatsubo, where Paravicini could recover while they waited for Yokohama to become habitable again.

However, we were not yet out of the woods. No sooner had we started chugging for the entrance than the *Steel Navigator* (whose Captain Simpson had been killed on shore) loomed up out of the smoke bearing down on us fast; and again we had to dodge frantically out of the way while she churned past with only a few yards to spare. By this time we had had quite enough of being chivvied about, by one thing or another, and drew a long breath when we too slipped out of the entrance into wide, choppy water.

Both lighthouses had been badly mauled by the earthquake and were down almost under water, the breakwater for a quarter of a mile on either side having subsided twelve to fifteen feet. In fact, the basin of the harbour and bay is reported to have sunk as much as eighteen feet. I can't vouch for this; but as the whole depression of Tokyo Bay is supposed to have been a huge crater when the world was young, and Professor Milne, the noted seismologist, many years ago predicted just such an earthquake as we were experiencing, the report of subsidences in the bay and off Kamakura are quite credible. It is also reported that the harbour of Misaki, at the end of the Yoko-suka Peninsula, sank twenty-four feet, submerging a couple of small islands that graced it and swamping the fishing villages around its edges; but during the next thirty-six hours the subsidence was reversed and the harbour restored to its normal level. There was no tidal wave within Tokyo Bay, and the only one outside seems to have been the comparatively small one along the beaches of Sagami Bay, including Kamakura, Dzushi and Hayama.

At the Commodore's request, the *Azuma* dropped us in a bight under Juniten, a small detached headland beyond the end of the Bluff whence a mile of delightful swimming beach sweeps around to the Honmoku cliffs. Here the *Daimyo* might ride in safety, protected from all but the most violent storms, until such time as Yokohama harbour should again be free of hazards. Left to ourselves, we made a hasty meal, battened down everything, and, tumbling into the dinghy, rowed out to the *Empress* to rejoin our undoubtedly anxious family. High time too, because in the course of a tremulous welcome after the terrifying strain they had been through, we learned that it was not Mabel Fraser who was on board the *Dongola* but a different Mrs Fraser. This was a jolt and called for prompt action. So, again manning the dinghy, with Ken Kruger to pull an extra oar, we rowed two miles over to the *Dongola* where, to our vast relief, we did find Auntie Mabel as well as the other unknown Mrs Fraser. Overjoyed to see us and discover we were all safe, she accompanied us back to a heartfelt reunion.

Her escape had been miraculous. Caught by the quake in her hired auto between Sakuragi-cho and Yokohama stations, and witnessing the terrible devastation on all sides, the chauffeur tried to drive back to the Settlement but got no further than the open square in front of Sakuragicho station, because a six-foot subsidence of the ground had left the two connecting bridges elevated in mid-air and inaccessible. Fissures gaped in the ground and fire was leaping up on all sides. All they could do was stay put and hope the surrounding canals would protect them. The chauffeur left her to try to get through to his family, but was beaten back. Hundreds of Japanese crammed into the small open space and crouched together to shield themselves from the furious heat as the fire encircled them. The wonder is they were not roasted alive.

Around 8 p.m. some young Continental, unknown to Aunt Mabel, fighting his way down from Tokyo to his family on the Bluff, came upon her alone in the mob, and together they crept along the shattered banks of canals, he carrying her small dressing case, eventually reaching Camp Hill where he parted from her to seek his family while she got out onto the reclaimed ground. Nearly all foreigners had been taken off by that time, but an officer from the last *Dongola* lifeboat, searching through the throng and shouting for any foreigners left, came upon her and she was rescued, scorched but otherwise unhurt.

By now it was six o'clock and growing dark, so, making fast the dinghy to the foot of the gangway by a long painter, we left it there for the night. Unfortunately, it was swept away in the darkness when an Anglo-Saxon Petroleum tanker *Iris*, commanded by Captain Koenings, daringly came to tow around the bow of the *Empress* so that she might steam out through the other ships and get away from the scum of oil now floating out through the harbour entrance, only half a mile away, which was burning fiercely right up to the breakwater and might at any moment catch fire outside. By this mishap to our dinghy, we were marooned on board like everyone else.

Even a refreshing wash-up did little to improve our disreputable appearance, since we were still in the same clothes in which we had waded through mud and scrambled down cliffs the day before; but we could not have cared less. The call to dinner sounded and we queued up hungrily where it was being doled out. With the children fed, we made some attempt to settle the family for the night, but it was so unbearably hot down below that we brought the boys' mattresses to the lounge, where they soon fell asleep.

Strolling about from group to group, we gradually picked up more news of the dreadful losses in our close community. John Mollison and his wife had been trapped in the fall of his father's office and could not be extricated before the fire reached them; whereas his eighty-year-old father, one of Yokohama's earliest residents, sitting in the next room, who would gladly have gone instead of his son, escaped. Their three small children were safe at Kamakura with their Scotch nurse, and were later sent home to England.

One heard, too, of the grit of Mrs Purington's nine-year-old son, who had been injured and his father killed in the fall of No. 168 Settlement. The father died with a parting injunction to their Chinese cook to try to save the boy. The cook carried him pick-a-back to the reclaimed ground, where the plucky boy lay with one mangled arm, begging helpers to leave him and take care of others because he was going to die anyway. When carried to a lifeboat, it was he who directed them how to lift him and how to set him down, all without a tear. He died on the way to Kobe.

Half the staff of the British Consulate had been killed in the collapse of the old stone building, a relic of earliest days. Boulter leapt out towards the tennis court and escaped with a gashed head, doing splendid work in the following days. Those who sprang for the front porch were mostly killed, including Hugh Horne, Haigh, Waddell and Lees. Grace Horne and her daughter Itta, who had been with Hugh, instantly dashed through the open French windows into the front garden, as did young Will Davies, the building falling about their heels. Hugh, only a pace or two behind, was overwhelmed. Besides being an able Commericial Counsellor, he was a wonderful gifted musician and a very kindly and humorous man.

The rickety old American Consulate likewise came down, burying the Consul Max Kirjassoff and his wife, as well as Miss Babbitt, Paul Jenks, Captain Simpson of the *Steel Navigator* and several others of the Consular staff. Kirjassoff was extricated and, with the aid of Monty Tipler of the nearby Chartered Bank, managed to dig out his wife; but by then the flames had taken hold and the others were lost. Tipler urged them to head with him for the Park, but Max seemed to think this would be going into the teeth of the conflagration and chose to head for the water-front, helping his wife along. Their bodies were found at the end of the road by the Custom House, where the fire overtook and trapped them, whereas Tipler survived in the Park. Witty and wizened Paul Jenks had been a fixture in the Consulate for many years and in him the community lost a rare character.

What was causing much amusement was the exploit of two young fellows of the Hongkong and Shanghai Bank: Jock Caldwell, a stocky rugged, good-natured man with a hard head, and Alan Guiness, tall and athletic, who sported a monocle and did it very well. Early on the morning after, they had gone ashore to take a look at what remained of the Bank, having the day before shared with indifferent success, in trying to lock up the valuables. The outer door of their safe would not close fast; only the quarter-inch inner door would do so. The ruins were still hotly smoking, but to their astonishment at their approach one or two figures darted guiltily away from the safe with objects clutched in their hands. On the run,

Jock and Alan nipped across the hot embers and found these looters had already pried away one corner of the inner door and rifled a couple of small compartments.

They faced a dilemma. If they went back to get a guard, the looters would return, prise off the door and get away with the entire contents. If one left and the other stayed, he would likely be scuppered. There was no way to secure the door. There was only one thing to be done: turn looters themselves and carry the contents of the big safe back to the *Empress*. So they finished the job of prying off the inner door, found a couple of iron rods in the debris, made a litter of their jackets and filled it up. This took them only to Compartment K. Then they used their shirts, which made excellent bags with the sleeves knotted. These brought them only to P. 'Come on', says Jock, 'off with your trousers', and two more clumsy bags were piled on the litter, but they were still only at S.

So off came their underwear, Guiness wincing modestly but still game. That did the trick, and, naked as nature made them, away they went down the pier with about two million yen in bonds and securities on the litter between them, a tricky pilgrimage over the partly swamped girders. As they neared the *Empress* heads began to crane and eyes to pop. What could this be, an enormously stout corpse carried by two naked men, one slightly overdressed in a monocle.

On they came. The officer on the bridge trained his glasses on the startling sight, noted that the two figures were white, not coppery, and ordered a relief squad down the gangway at the double, flourishing jackets, coats, trousers, etc. By this time Jock and Alan were dead to shame and almost brushed them aside, but consented, for the sake of blushing women on board, to having a few robes thrown around them. But relinquish their burden, never! They toted it aboard and wouldn't let it go until it was safely in the ship's strong-room. They were the day's heroes and had thoroughly enjoyed themselves!*

At about eight o'clock that Sunday evening, Boulter, Acting British Consul, supported by the Reverend Eustace Strong, summoned all men aboard to a meeting in the card room to organize rescue squads for the following day, to search for missing people and any survivors in the outskirts. In view of the violence and bloodshed on shore, only bachelors would be allowed to volunteer. It was no fanciful risk. Bill Blatch of the Rising Sun (Shell Oil)—a young friend of mine who had done a ten-day mountain trip with me three years earlier—was attacked by a mob in Kamakura village, mistaking him for a Korean suspect, and was being bludgeoned to death when saved by the timely appearance of a Japanese cavalry officer. Married men were asked to remain on board and see to getting their families away on ships which were already converging

---

*I set down the story of this incident just as it was joyously circulated on board the *Empress of Australia* the next day. Long afterwards my friend Charles Rice assured me that it was not Alan Guiness who had helped Jock but some stranger. Though loath to relinquish that cherished vision of Guiness triumphantly striding down the pier attired in nothing but a monocle, I decided to seek the truth.

In 1960 I wrote to Jock Caldwell, retired in Sussex, England, to ask what actually happened. His reply shows how gleefully the grapevine sometimes strays from the narrow path of veracity. He wrote:

'I am sorry to say, being reluctant to spoil your good story, that Charles Rice was correct. I did collect a passing stranger—a beachcomber, in fact—on the Monday morning when I eventually got ashore from the *Andre Lebon* to look for Stinie Morrison. This man and I emptied the contents of the Accountant's safe, bonds, share certificates, title-deeds, &c.—into our respective shirts, singlets—and my trousers. I still wore a *fundoshi* (loin-cloth) however. I got the "bodies" onto a launch operated by Captain Heseltine at the Canal Bridge by the Grand Hotel and we took the booty out to the *Empress of Canada*, which had just arrived from Vancouver on her regular run and was anchored in the Bay. A funny incident occurred at the

upon Yokohama and would be arriving next day. The *Empress of Canada*, approaching Japan from Vancouver, had wirelessed that she would be in at 9 a.m. and would take as many refugees as she could accommodate to Kobe; and any who wished to go further would be carried on to Shanghai or Hongkong. The *President Jefferson,* on her regular run from China and Japan to Seattle, would be arriving from Shimizu, the port of Shidzuoka half way down to Kobe, at daylight on Monday, and had wirelessed that she would change her routing and take a full load of refugees to Kobe, proceeding thence to Seattle; and any who wanted to carry on to America would be free to remain on board.

The *Empress of Australia* herself was, of course, still unable to put to sea until divers could be procured to free her propeller.

This was the situation when we were finally able to turn in that second night.

ship's gangway. I, very dirty and still wearing only the *fundosh* and a *tenugui* (cotton cloth towel) round my forehead, jumped from the launch onto the flat foot of the gangplank, with a view to helping a woman off the launch first, whereupon the Canadian Quarter-master socked me on the jaw and over I went into the water. I came up spluttering and called him many names he was not baptized with. He quickly pulled me up and apologised, explaining "Sorry, Sir, I thought you were a coolie!"

'The securities were safely taken aboard and, I was told, were insured by wireless for ten million yen (US $5,000,000) for their subsequent voyage. Guiness was definitely not with me during this episode. (You'll be sorry to hear he died early last year, only sixty-four.) I parted with him and the others on the hatoba that awful Saturday afternoon and got aboard the *Andre Lebon,* and then had the extra terrifying experience of nearly being burned up when the oil in harbour caught fire on Sunday morning. The *Lebon's* engines were out of order and the Captain let out cable; but it was Tom Laffin with a launch and a strong line that got us out to a distant buoy in the harbour and we escaped the flames which raced past us. He got the Legion d'Honneur and a monetary award from the French Government, and deserved it.'

✄ IV ✄

EXODUS

Early next morning, Monday the 3rd, we were told by the ship's officers that we were to be transferred to other ships, that we must decide where we wanted to go and take designated gangway stations accordingly.

So far we were still under the illusion that all Japan had suffered equally from the earthquake and that our sister port Kobe was in sore straits too. Concluding that, if this were so, life for the foreigner in Japan had virtually come to an end, I had during the night made up my mind to send Dorothy and the boys by the *Jefferson* to Seattle, there to mark time until we could see what lay ahead. I would accompany them as far as Kobe, remaining behind, if possible to see the situation through. Mr Campbell had to remain in Yokohama aboard the *Empress of Australia,* to attend to a Pacific Mail Ship arriving that week. Mrs Campbell would go along with us to Kobe and await him there. Mabel Fraser wanted to go back to England.

But, as with a twist of a kaleidoscope, the whole picture suddenly changed. While waiting with Dorothy and the children in the crowded maindeck, lined up for transfer to the *President Jefferson,* Mrs Campbell fainted from the stifling heat and we had to carry her up to the promenade deck, relinquishing our places in line. While reviving her, Tom Laffin, (still doing splendid work with his launch) brought me word that a Japanese

14. A burial party in the new foreign cemetery near the Y.C. & A.C. grounds in the hills.

F. W. Horne's striking residence, 'Temple Court', at No. 10 Bluff, skidded down into Jizo-zaka hill. The temple whose priests helped to design it stood in hollow on left. Japanese part of Yokohama in distance, temporary houses already springing up.

Gate to the Customs Compound with the walls of Custom House beside it on right. Pier already functioning again.

15. The Come-back. W. W. Campbell hoists the Pacific Mail flag and reopens his office in the vault of the old building at No. 21 Settlement.

courier from my father's office in Shidzuoka had come up by the *President Jefferson* with a message that father was safe and well, that Shidzuoka had been but moderately damaged, and that father anxiously awaited news of us. With what relief and gladness I sent back assurances that we, too, were safe and that I would communicate with him from Kobe.

Simultaneously, someone announced that R.F. (Bob) Moss, who had been up-country at Karuizawa, had got down by rail to within fifty miles of Tokyo and struggled through the rest of the way on his motor-cycle, bringing word that Karuizawa had suffered, but not too seriously; that Tokyo was two-thirds destroyed by fire, but that the earthquake damage to the un-burnt portion was nothing like as bad as in Yokohama. That meant that my sister and family in Karuizawa were certainly safe and would be heading back to Shanghai as soon as communications were restored. Now I could draw a long breath.

So after all Japan was not wiped out, and the area of the cata-strophe did not extend beyond perhaps one hundred miles in all directions from Yokohama. (The epicenter of the earth-quake was later found to have been the bed of Sagami Bay, about twenty-five miles west of Yokohama, off Kamakura). This news acted like the sun coming out from behind a cloud. There and then we switched over to the gangway for the *Empress of Canada,* by which Dorothy and the boys could go on to Shang-hai where I knew my brother-in-law, N. G. (George) Maitland, and later Eleanor, would gladly harbour them in their large residence until I could bring them back to Japan.

Mabel Fraser was with us, of course. Also Mrs Wheeler, grief-stricken over the death of the old doctor and exhausted from two days alone with her faithful servants in the stables of 97 Bluff, where she had taken refuge when the house came down and burned. Mabel had already been ministering to her and continued to take care of her on the way to Shanghai, later accompanying her all the way to London to join her daughter May Murray, one of Mabel's girlhood friends.

Frank Anderson, Honor and Patsy, who had stayed with us throughout, were also transferring to the *Empress of Canada,* he to drop off at Kobe and Honor and Patsy to go on to Shanghai. Bateman, Gladys and Mrs Syme Thomson chose the *President Jefferson,* the ladies transferring at Kobe to the *Empress of Canada* in order to continue to Shanghai. Of the rest of Dodwell's staff, I ascertained that George Colton, Gordon Bell and Mary Martin were aboard the *President Jefferson,* Lucy Fox on the P. & O. *Dongola.* All landed at Kobe except Miss Martin, who went on to the States. Not long after arrival, however, Gordon Bell threw in his hand in order to escort a young American girl of his acquaintance whose mother had been killed, back to California. How Jeffery made his way down to Kobe, I don't recall; but he reported there for duty and our whole foreign staff was thus accounted for.

During that Monday morning, the *Empress of Australia* disgorged her hundreds of refugees, including most of the injured, into lifeboats for transfer to other ships. Still as grubby and ragged as ship-wrecked emigrants, two boatloads of us were towed to the *Empress of Canada,* the husky sailors finding some difficulty in lifting the disabled onto the slender gangway. Doro declares I saved the day for her as she ascended the steps with tears in her heart, by calling out cheerily from below 'Thank the Lord not thirty-two pieces of luggage this trip!' Only the previous year we had gone on home leave, my first since bachelor days, and to my consternation I found I had accumulated a wife, three children, an English Nanny, and thirty-two pieces of luggage! Now we were off again on an unknown voyage, with all our worldly goods tied up in my pocket handkerchief slung over one finger—the children's underwear which

Doro had washed in our cabin that morning and was not yet dry.

Old Miné, the children's devoted amah, was still with us, her seamed face as placid as an ivory *netsuke*, accepting all the buffetting with benign stoicism, unconcerned about the future as long as she could remain under our wing. On all sides one heard of the loyal devotion of Japanese servants and staffs who without exception behaved heroically in rescuing and standing by their employers, often at great risk to themselves. In fact, the way the Japanese individually met the disaster, and uncomplainingly started afresh from scratch, was admirable.

Once aboard the *Empress of Canada,* we were given a nice stateroom and almost literally taken into the arms of the sympathetic passengers from America who, staggered at the enormity of the disaster and the condition of the refugees, opened their trunks and gave out clothing until they had little left for themselves. Frank Shea, seeing how muddied and torn I was, insisted on giving me one of his light grey suits, which proved a godsend as the Kobe outfitters were stripped bare before I reached them; and I had to live in borrowed clothes until I could get down to Shanghai and fit out afresh.

The same kind help was given on other ships and it was told that a good-hearted quarter-master, observing that N. F. Smith was soaking wet, rigged him out in one of his own sailor suits, and the old gentleman, pleased as Punch, tottered around the decks, side-whiskers streaming in the breeze, weaving from side to side as the ship rolled, for all the world like a character in *H.M.S. Pinafore* doing a sailor's hornpipe, and quite unconscious of the sensation he was creating! What a contrast to one of my boyhood memories of him, on my very first shooting trip, sloshing ruggedly through the Yawata marshes and bowling over snipe with unhurried precision.

Helen and Van Smith had elected, like ourselves, to transfer to the *Empress of Canada* and were in a cabin close to ours, with little Maureen growing steadily worse. The doctors had decided that only an operation could save her and it was to be performed that afternoon.

Morris Mendelson also came aboard, still in a daze from grief and shock. Yet another friend of our boyhood turned up too: Charlie Thorn, who had succeeded to his father's publishing business, the 'Box of Curios Press', started in the pioneer days. He was even better known as a tennis player and leader of the Bijou Orchestra. He had escaped from his business premises by sheer agility, hurdling desks as the building collapsed behind him, trapping a number of his Japanese craftsmen. It was all a knock-out blow as the business went with the plant and he had to start life afresh in one of the large American merchant houses. His wife Muriel and their two boys were fortunately up country and came through safely.

All through Monday the *Canada* lay at anchor, expecting hourly to sail, but lingering on for further groups of refugees coming aboard at intervals from the shore.

The *President Jefferson* and the *Dongola* both got away that afternoon, reaching Kobe next day, Tuesday the 4th. On the way down to Kobe, the Captain of the *Jefferson* offered to transmit messages by wireless for any of the passengers. This could be done at sea, whereas vessels in port were by regulations not permitted to despatch messages; hence the scant news that reached the outer world. Bateman thoughtfully sent one to our Hongkong office telling them that our Yokohama staff, their families and Mrs Syme Thomson (who was the widow of our former Manager in Yokohama) were all safe and proceeding to Kobe; but that offices, godowns and residences were completely destroyed. Hongkong at once passed the word on to London Head Office and this was the first message to reach

London from the disaster area. Its prompt dissemination by our Head Office among other Far Eastern firms brought the first ray of hope to many who feared that the whole population of Yokohama had perished.

Among the last stragglers to come aboard after dusk was Miss Lauritsen, the pretty Danish girl whom my wife had brought out with her from England in April as governness for our children. She was a bright girl and brought Tony on quickly, but was actually far too attractive for the quiet role of governess, with the result that just before the earthquake, to our mingled disappointment and relief, she left us to become a stenographer with my old school-mate Willie Squire. Before starting work, she was taking a good holiday at Miyanoshita, and the quake caught her there. With admirable spirit and grit, she next day joined four men in an attempt to get down the pass on foot, a dangerous feat, as the narrow road had fallen into the gorge in many places; but they succeeded in reaching Kodzu, seventeen miles away on the railroad, by nightfall. Of course, the railroad was destroyed and they spent the night in an abandoned automobile. Next day a station hand miraculously obtained bicycles for them and they pedalled the thirty-three miles to Yokohama through a devastated country-side. At Kanagawa the men pushed on to Tokyo, she finishing the last four miles of ruins alone, and being finally picked up at dusk by Dave Abbey and Morris wandering along the deserted Bluff Road, the solitary living soul in that wilderness. Grate-fully, she accepted their escort to the lifeboats. Next day, she appeared on deck in a long chair, full of vitality, holding court to a group of young admirers. She said the guests at Fujiya Hotel had escaped without loss of life and were camping out on a ledge with scant shelter but an ample food supply. This was reassuring news for many with friends up there. Miss Lauritsen travelled in the *Empress of Canada* to Shanghai, and we were pleased to hear the following year that she had married a well-to-do resident.

Early on Tuesday morning, September 4th, we sailed for Kobe without another chance to get ashore. During the morning, the passengers from America held a meeting in the saloon to take relief measures and collect subscriptions. Frank Shea, General Manager for Japan of the American Trading Co., of Tokyo, and I were asked to address the assemblage, he for Tokyo and I for Yokohama. Shea, however, persuaded Mac-Dowell of the American Embassy to take his place, feeling that as he himself had been aboard the *Empress of Australia* during the quake, he was not competent to describe the situation in Tokyo, whereas MacDowell had been on the spot. It was a moving service, with hymns and prayers for the dead; and at the end of my descriptive talk, when I tried to thank them for their sympathy and aid, I was embarrassed to find myself choking up. They were wonderfully generous in raising several thousand dollars for the destitute.

While the meeting was going on, all refugees on deck were asked to remain quietly in their places so that a complete list of survivors could be made. The result was broadcast by radio and we in turn received lists from other ships, so that gradually it became known who had been saved and where they were. Unfortunately for me, having been at the meeting below, my name was not recorded, only 'Mrs O. M. Poole and three children', so that many people, including my brother Herbert in charge of the Standard Oil Co. in Mukden, Manchuria, concluded I had been killed and only my family saved.

That afternoon, on the way down the Japanese coast, little Helen Maureen Smith died; and at sunset we buried her at sea from the stern of the ship, two of the ship's officers in white uniforms conducting the simple service. It was a touching scene in the tranquil, lavender light, with Van bowed in grief and Helen on my arm. A tragic sequel to their marriage of only a short year.

Soon after daybreak next day, Wednesday the 5th, we arrived at Kobe, a bright, sparkling Kobe unmarred by the quake, with its familiar steep green hills behind. Although nothing had been known there of the real nature of the disaster until the first ship, the *Philoctetes,* steamed in on Monday, the 3rd, the foreign community then sprang instantly into action, and before the first heavily laden refugee ships arrived next day, Kobe had been organized as one huge relief camp, everybody throwing open his home. Within hours, business men had arranged to send up a relief ship, the *West Orowa,* which left at midnight with medical supplies, provisions and drinking water, as well as a rescue contingent of a dozen foreigners representing the community, among them James H. Ewing, a promising young recruit to Dodwells Kobe office, out barely a year from London and eager to lend a hand in this emergency. The *West Orowa* arrived at Yokohama at daybreak on Wednesday, the 5th and though by that time the bulk of the fleeing community was already afloat, her aid was very welcome and she sailed for Kobe next day with fifty of the last to leave Yokohama, arriving on Saturday, the 7th. A second trip followed in a few days, with specially requisitioned supplies for the remnants of Yokohama's community clinging tenaciously to what was left of their former thriving city.

Kobe was inundated with refugees and everyone was doubled up. Tireless committees generously gave out clothes and necessities to all in distress and naturally the stores were soon denuded of everything. Subscription lists were opened and funds poured in as never before, enabling many acts of kindness to be performed instantly. I can't begin to tell what these generous people did; it was beyond praise.

E. A. G. May, Manager of our Kobe branch, met the

Empress of Canada on arrival and kindly took Mrs Campbell and me into his home at San-bon-matsu on Kitano-cho, high up on the hill overlooking the city, where he was living with his charming young wife Doris and their baby. Dorothy and the three boys remained on board the ship, which sailed for Shanghai that night, taking also Gladys Bateman and Mrs Syme Thompson, Honor and Patsy Anderson, Mabel Fraser and Mrs Wheeler, among the hundreds of other refugees leaving Japan temporarily or for ever.

Next day, Thursday the 6th, I had a cable from George Maitland to say he was on the way from Shanghai to rescue Eleanor from Karuizawa where, though safe, the summer colony was isolated and experiencing some hardship from extensive damage to the hotel, village, and summer cottages dotting the plateau between encircling mountains. Their casualties, fortunately, had been light. 'N.G.' was not, therefore, in Shanghai to meet Dorothy and the children when they arrived. However, Major Hilton-Johnson met the ship, George Moss of the British Consulate (Hilare Tait's brother, later Sir George Moss) took them ashore, and Dorothy and Ralph Melhuish (our close friends in the Hongkong Bank), Brand (George Maitland's partner) and Hugh Lester with all Dodwell's people, met the party on shore and took them into their homes, Ralph coming to sleep at the Maitland's house every night until N. G. and Eleanor returned.

My brother Bert, representing the Standard Oil Co. in Mukden, Manchuria, unable to stand the suspense of not knowing what had happened to me, hopped on a train on Wednesday the 5th, arriving in Kobe on Saturday, determined to search for me and take care of Doro and the boys if need be. Greatly relieved to find me alive, and reassured about Eleanor and the Dad, he returned to Mukden that night.

That same weekend the refugees from Miyanoshita arrived

in Kobe by rail from Numadzu, including many of our friends: Edna and Percy Brown and their small boys Norman and Michael, Alice Mendelson, the H. A. Stewarts, Georgie Reynell and her two children, the C. L. Spences and perhaps twenty others, informally led by Captain David James, who did splendid work in organizing the exodus. Far-famed Fujiya Hotel at Miyanoshita had remained upright, though so badly damaged that it was no longer habitable; and the guests, who had all escaped with only two injured, camped out for the next four days on the ledge of the tennis court behind and above the hotel, terrified by continuous shocks and the frequent roar of landslides down the precipitous slopes of the mountains surrounding them, and being made miserable by deluges of rain. The village, clinging to the edge of the gorge below the hotel, had hurtled at both ends into the chasm, carrying a quarter of the villagers to their deaths, including one American, a Mr Herr. No relief could reach Miyanoshita as the winding road up the gorge from Yumoto had sloughed away in dozens of places, as had the upper mountain road over Otome-toge to Gotenba.

On the fifth day, after suffering many alarms and hardships, the desperate guests, still without news from Yokohama, trekked in a body, afoot or on horseback twenty-two miles over the mountains past Hakone Lake and down to Mishima, where they were able to get automobiles to Numadzu on the seacoast and thence by the patched-up railway to Kobe, where for the first time they learned what had actually happened to Yokohama, their friends and families. Both the villages on Hakone Lake and the two hotels for foreigners had suffered terribly, being almost totally demolished; a young English couple, the Tebbuts of Shell Oil, were killed. The husband was overwhelmed and never seen again; his wife was pinned from the waist down, and though she had lived through the afternoon, death came as they lifted away the heavy timbers that held her. Their baby escaped and was sent home to Tebutt's people.

Back in Yokohama, relief was pouring in fast. The first to arrive were four American destroyers on Wednesday the 5th, which had raced at full speed from Dairen, Manchuria, immediately on hearing of the catastrophe. Brushing aside all restrictive regulations, they steamed right up to Tokyo to give immediate aid to the diplomatic corps and, of course, to Yokohama sufferers. On Thursday, H.M.S. *Despatch* arrived from Shanghai, followed by the American cruiser *Huron* on the 7th and the British *Hawkins* on the 10th. Merchant ships were also turning up from all quarters and by the end of the first week virtually all foreign refugees had been taken off, leaving only a few hardy souls so inseparably linked to Yokohama that they were determined to stick it out as best they might.

The *Empress of Australia*, freed at last from the *Steel Navigator*'s cable by divers from the Japanese battleship *Yamashiro Kan*, which had arrived from Kuré Naval Arsenal on the 4th, was again seaworthy by the following day the 5th, but her Commander, Captain Robinson, and the British Consul decided to keep her in port a few days longer to care for refugees still straggling in from round about. She finally sailed at midnight on Saturday the 8th, arriving at Kobe Monday morning with about 600 foreigners and several hundred Chinese on board. Of the hundreds of Japanese who had taken refuge on her from the first fury of the quake and fire, only the seriously wounded remained, the rest having gradually filtered back ashore to their own people.

Among the arrivals by the *Empress of Australia* was my popular father-in-law W. W. Campbell, who gratefully accepted May's invitation to join us, as did Fred. W. Hill, Winnie Rice's father. A week or so later, the Pacific Mail Agent was transferred elsewhere and the Campbells moved into his vacated residence. In view of the general congestion, May decided to send his wife Doris and their baby over to her mother in Vancouver, to clear the way for making a Dodwell Mess of the Manager's house. Before long, therefore, I found myself presiding over the 'San-bon-matsu' ('Three Pines') Mess, consisting of May, Bert Bateman, Frank Anderson, E. R. (Bob) Hill from our London office, who had arrived in Japan two days after the quake. It was a lively group and our mess life reminiscent of care-free bachelor days.

PHOENIX

For me those first days in Kobe were filled with anxious hours delving into the Company's position with our re-united staff, as well as constant community and committee meetings. As a member of the committee of the Yokohama Foreign Board of Trade, I was automatically drafted into the joint Kobe and Yokohama Relief Committees, in almost continuous session over emergency measures: arduous and sometimes saddening work, with so many interests involved in every issue. But it was stimulating to see what could be accomplished in the face of difficulties by the resourcefulness of a score of determined men working together.

A few days after reassembling in Kobe there came an opportunity to send one of our staff back to Yokohama by ship and, my presence being required in Kobe, Bert Bateman volunteered to go. He was back in three or four days with confirmation that nothing on our premises had survived but my own Chubb safe, which had been dynamited by a squad of British sailors from a destroyer and the contents placed on board in a sealed sack. He found some of our Japanese staff hovering around the ruins, all of whom expressed willingness to transfer to Kobe if summoned. Bateman also visited the site of his house to see if anything was left, and came upon a remarkable sight: a dozen dogs, pets of owners forced to flee without them, who had formed themselves into a comically-assorted wolf-pack and were roaming the Bluff in search of food. Among them was his own golden-haired Pekinese pug, last seen at the edge of the cliff in the British Naval Hospital. With all the beautiful 'feather' burned off her tail and hind quarters, she was intrepidly ranging shoulder to shoulder with other 'wolves' twice her size. At Bateman's whistle, she whirled unbelievingly, made a rush and leapt straight from the ground onto his shoulders. Needless to say he brought her back with him to the mess.

On Bateman's return, I managed to get back to Yokohama myself for a few days to conclude arrangements with our Japanese staff as to where and how to carry on and to consider on the spot the advisability of establishing our main office in Tokyo, maintaining only a shipping office at Yokohama. The trend of big business towards Tokyo had long been perceptible, and now that Yokohama as a complete and efficient trading mechanism had ceased to exist, it was inevitable that Tokyo, with its enormous population and resources, should become the centre for commercial dealings.

The blackened ruins of Yokohama were a harrowing sight. One or two roads in the Settlement had been partly cleared, but the rest were so blocked that it was hard to distinguish their course. And everywhere lay charred corpses, pathetic shrunken mummies. The canals, docks and slips were filled with bodies, bloated to an orange shapelessness, scattered amongst masses of rubbish that choked every waterway. Here and there the funnels of sunken launches protruded from the scum. Many of the bodies of Japanese which still littered the streets were being methodically collected into piles of four to six and crudely, though with surprising completeness, cremated under sheets of corrugated iron laid across them. The smell of burning was bad enough, but as one edged through the sea of ruins the putrid odour of death became so pervasive that even a eucalyptus mask could not shut it out. Weeks were to pass before the debris could be cleared and decent burial be given to remains found underneath.

A few of the most substantial edifices had defied the earthquake and stood like monuments among the fallen rubble. The Russo-Asiatic Bank's new steel and concrete building was externally intact though gutted by fire. I climbed its wide marble staircase to the second floor where, incongruously, lay the carcas of a horse. The poor abandoned creature must have climbed the stairs in a last attempt to escape the flames. Similarly, the granite building of the Yokohama Specie Bank on Honcho-dori was also standing and here a terrible tragedy had occurred. In some manner, its fire-proof steel portals had been closed upon a number of its staff, patrons and passers-by who considered it a safe refuge and; unable to get out again when the fire came, they were suffocated in heaps.

Of our own office there was nothing left but piles of baked rubbish. Even the location of our rooms could only be determined by chimneys and fireplaces. So flattened was the Settlement around us that I could see the ships in harbour from our front steps. By previous arrangement our head Japanese were on the site to meet me and with picks and shovels we sought anything that might help in our task of reconstruction; but the fire had done its work too well. It was fruitless.

After parting with our Japanese, I made my way through the Settlement and up Camp Hill to the Bluff, taking photographs which still harrow me. There was no longer any smoke and in the clear atmosphere the ashes of Yokohama stretched to the distant hills, Lean-tos were being thrown up here and there in the Japanese city, but the foreign Settlement and Bluff were still deserted areas of black desolation. Our house at 68 Bluff was just an ugly scar and depressed me horribly. Bowden's, Hayward's and Campbell's houses provided grotesque examples of how badly structures could be battered without coming down. A. P. Scott's house, adjoining the Bluff Gardens, was completely pancaked, capped by its more or less intact roof like a pie-crust. A fissure traversed the lawn.

The fleets, American and British, were doing fine work, demolition squads cracking safes, landing parties bringing in refugees from outlying places, still others digging out bodies and burying them temporarily, some in the British Consular grounds, others in the Royal Naval Hospital close to where they were found, and others out at the new foreign cemetery in the hills near the Y.C. & A.C. Grounds. One misty morning I attended the burial of two old friends out there, one of whom had finally been identified by an entry in a vest pocket engagement book that had not been burned: 'Thursday 6th, dinner with Poole'. I knew who the six of us were to have been. Two had been killed and recognized; two had escaped; this could be the one man missing, Jock Watson. But they had already buried elsewhere someone thought to be him. After a re-check, the other body was identified as another Club member, wearing a similar pepper and salt jacket.

As for Tokyo, my first sight of it was appalling. Not that its standing ruins were anything like as battered as those of Yokohama, but the immense ocean of flattened, burned-out city beggared description. On the other hand, the modern business section, confined to an area of a few blocks in Marunouchi, had not suffered critically, its steel and concrete office buildings having withstood the shocks, with a few twisted exceptions; and though some had been gutted by fire, the remainder could be made operative again without rebuilding. The outstanding case of survival was the Imperial Hotel, built by the American architect Frank Lloyd Wright, who, because of the spongy nature of the ground, erected his Aztec masterpiece on floating foundations. Except for a two-foot subsidence of the dining room building in the middle of a central court, all the buildings came through intact. Needless to say, its survival proved a haven to many.

Tokyo's tragedy was not the earthquake but the ensuing fire which spread so fast that thousands were overtaken and burned to death. As the populace fled, hordes took refuge in the open grounds of the Army Clothing Stores, as large as a parade ground, believing the fire would pass them by. All carried bundles of clothing, quilts, etc. and some dragged handcarts piled high. The fire was actually passing them on both sides when a sudden change of wind bore it down on their flank, simultaneously cutting off all exits. The mounds of bedding and salvage quickly caught fire and 32,000 people were roasted to death that night in the confined space. I saw some large photographs of the scene taken next morning (and quickly suppressed) which were terrible beyond words, bodies touching each other as far as the eye could see. The victims were all cremated on the spot and a monument erected there of cement made from the ashes of the dead. A sombre yet fitting memorial.

I am not sure of my facts but to the best of my belief no foreigners lost their lives in Tokyo. Many had narrow escapes and experienced hardships, losing their homes in various parts of the far-flung city. Those who worked in Tokyo but lived in Yokohama had a perilous time battling their way down the eighteen miles to find their families. The roads became more and more impassible as they came closer, so that only a handful accomplished the journey before dark and most had to trudge through the string of villages along the Tokaido at night with only the glare of fires to light the way, arriving at dawn to find the seaport completely destroyed and the Bluff a wasteland.

Our company quarters in Tokyo were in one of the smaller modern buildings near the Tokyo railway station and, although seared by fire, proved to be still sufficiently habitable to serve while looking for larger offices. Meantime, our key Japanese were to come to Kobe to help in tackling the formidable task of discovering where we stood. I need not dwell on this phase. Every firm had its individual problems which seemed to breed their own solutions when persistently tackled. We on the spot had naturally to bear the impact of the earthquake in its varying forms, but Head Offices in New York, London and elsewhere had their share of heavy anxieties and took prompt steps to meet the emergency. In the British Colony of Hongkong, Far Eastern headquarters for many firms (which had just been through a devastating typhoon with great loss of life), the physical aspects of the situation in Yokohama were instantly grasped and quick action taken to send succour. A former P. & O. liner was jointly chartered, renamed the *Tai Wei Fung*, fitted out as a floating hotel, heavily provisioned and despatched on her mission within ten days, arriving at Yokohama about a week later. There she tied up at the patched pier for a long stay and supplied the most pressing need of the moment—adequate living quarters in the midst of havoc, a boon to the few still remaining and to the many streaming back to cope with personal problems and re-establish former occupations. Each subscribing firm had permanent quarters aboard, Dodwells' being a roomy three-berth cabin. Captain C. Heseltine, of our stevedoring associates, F. Owston & Co., enjoyed one of our berths, doing excellent maritime work for us and proving indefatigable for others too. The other two berths we kept for our own staff, who were constantly coming and going in turn.

I was an early occupant and it was always with relief that I left the stinking, hot, dusty shore, took a cold shower and, garbed in Summer *yukata* (bath kimono), dropped into a deck-chair with a long whisky-soda and shook off the depressing murk of calamity saturating the ruins.

Like all ancient P. & O. boats, the *Tai Wei Fung* had her dining saloon in the stern, two long tables running down its full length. In the evening hour before dinner these tables became the 'Club Bar', where everyone gathered to exchange the latest news picked up on shore or on the radio. At the head of one table sat F. H. Bugbird, Manager of Jardine Matheson & Co. of Yokohama, a rugged old-timer of irrepressible good spirits, noted for a fabulous fund of anecdote. He never told the same yarn twice and could fashion any incident into a most hilarious story. Evening after evening he presided at his table and it is no exaggeration to say that his boisterous humour saved some of the hardest hit from despair. At the head of the other table sat big Edward Coutts, the exchange broker, his injured back put to rights, whose slow geniality provided the perfect foil for 'Buggins' sallies. On deck after supper, Jock Caldwell of the Hongkong Bank fell naturally into the role of entertainer, heading a young group in organizing sing-songs, card games, smoking concerts, etc. and seeing to it that no one was ever without a drink, all done with such spontaneous goodwill and hospitality that everyone was carried along on the wave of comraderie and seldom turned in before midnight. But for this nightly relaxation, the constant tension would have strained one's endurance. It is hard to picture, from a comfortable environment, the fantastic disorder that surrounded these men in their floating oasis.

I recall how impressed were the special representatives sent out by Lloyds of London by evidence of the blow-torch intensity of the heat generated by the wind-lashed conflagration. One evening they brought aboard from the Customs piers what remained of two kegs of nails taken from a stacked pile. Before the wooden kegs had had time to burn away, the nails had been fused tightly together into the shape of the keg. And in one case, where the keg had rested on the cement pier, its lower nails had fused into the cement so that there was no telling where nails ended and cement began. They were taking these trophies back to London to show why there was practically no salvage anywhere.

The ship-borne colony operated not only individually but as committees to tackle specific matters of common concern. On one occasion I accompanied four fellow-Americans to lay a pressing situation before the American Admiral aboard the U.S.S. *Huron*, which had arrived on the heels of the first flotilla of destroyers. He austerely agreed to receive only our senior member, Everett W. Frazer, leaving the rest of us to cool our heels—and our bottoms—on the steel deck under one of her guns, enviously watching her husky gobs tuck away a healthy meal at long deal tables. He did, however, accede to the object of our mission without hesitation, so we forgave him.

Meanwhile, signs of life were re-appearing on shore. Once the burnt-out embers had cooled off and the bodies of thousands of dead removed, flimsy wooden shelters sprang up like toadstools over the featureless ashes in Tokyo and in the Japanese sections of Yokohama; but the foreign Settlement and Bluff, denuded of life, remained barren and desolate except for a sprinkling of shacks to establish a *pied à terre*. Then, as lumber became freely procurable, more substantial sheds took shape, mainly in proximity to the pier. Behind the Grand Hotel corner emerged something called the 'Tent Hotel', a cluster of plank and canvas cabins providing simple shelter and food. Dodwell's Seattle office quickly shipped us some prefabricated bungalows, which we set up here and there, where most desired. One of several along the Bund served as a small office for our young shipping-man, John Barnett, with two Japanese assistants.

Except for these meagre stirrings, the Settlement remained for months a mass of deserted rubble, its torn walls standing

like tombstones over a dead past. Former property owners, by
no means certain that trade would return to the crippled port,
were loath to re-invest in buildings exposed to the hazards of
earthquake and political uncertainty. Costs had risen many
times since pioneer days whereas opportunity had dwindled
with the growth of Japanese competence in international com-
merce.

The worst blow to recovery came with the announcement
that losses incurred in the holocaust had been ruled irrecover-
able under fire insurance policies, in which the small print ex-
cluded 'fire from earthquake.' Hardly anyone had been aware
of this. Direct damage from earthquake had been so minor over
the years that an additional earthquake premium of one-half
per cent appeared unwarranted in face of the general assump-
tion that fire insurance covered all fire risks. Disillusionment
came as a staggering blow to the community of whose members
it seems that only four carried earthquake insurance: the
United Club, the Grand Hotel and, because of their art col-
lections, C. K. Marshall Martin and John R. Geary. The ma-
jority of residents, basking in false security, discovered they
had lost all their possessions without hope of redress. This
applied equally to merchant firms whose establishments and
merchandise had become a total loss. It was a shattering blow,
in many cases spelling ruin.

Kobe had become headquarters, not only for immediate relief measures but for everything to do with reconstruction. It
was now the only major foreign settlement in Japan and everyone congregated there. Having lived in Kobe for three years
during my bachelor days, I still had many old friends there
who were kindness itself. All members of the Yokohama United
Club became, by invitation, members of the Kobe Club, which
was thronged every noon and evening by a suddenly doubled
membership, eagerly discussing latest developments. Com-

mittees met daily in the airy second-floor rooms; and many differences were ironed out over a friendly drink in the bar.

The basic problem facing everyone was this: banks, merchants and individuals had lost not only their possessions but all their books and records too. Japanese customers also had lost their records, and some their lives. Everything had been consumed. What hope could there be of sorting out the mess?

Incredibly, the jig-saw puzzle was slowly put together. Personal accounts with banks were re-instated by matching the individual's recollection of his status with the phenomenal memories of the Portugese clerks employed by most foreign banks, and compromising any difference. Company accounts were naturally more involved; but it was fantastic how, by one means or another, positions were credibly reconstructed.

It was the common practice of banks and large merchant houses too send copies of their books to head office every month; and when our London Office sent back ours to us, we at least knew precisely where we stood on July 31st. But August was a complete blank and, except for some valuable documents in my safe, we had nothing to go upon.

Our first step was to assemble the Yokohama staff, foreign and Japanese, in Kobe and record our composite recollection of the overall position. Then we split up into departments and did the same thing in greater detail. We next held sessions with our Japanese buyers and suppliers and recorded all they could remember. Our godown staff, partly Chinese, proved invaluable in recalling what deliveries of merchandise had been made to customers or taken in; and our landing agents contributed pivotal information as to the status of lumber cargoes in process of delivery, a matter of great importance since the logs had either been burned in the customs slips or swept out into the harbour or bay and subsequently washed up on distant shores. If undelivered to buyers, the marine insurance still covered; if delivered, the risk had terminated. Marine insurance did not exclude fire from earthquake.

It was all intensely complex, but gradually one piece of evidence fitted into another until a discernible picture emerged with surprisingly few permanent blanks, and we were able to draw up a tentative statement to work upon persistently. I want here to pay a tribute to our Japanese merchants, who scrupulously admitted their liability when circumstances pointed to it, although evasion would not have been difficult. Their integrity was admirable.

After months of hard work, we finally knew where we stood and were in a position to carry on from where we had left off, with our prospective loss worked down to a tenth of the ghastly figure that at first had been at stake. That considerable residue we had to swallow with good grace, thankful that it was no worse.

Not all were as fortunate. Many who had lost everything lacked the capital to re-establish themselves and had perforce to join other surviving concerns. But being men of vigour and experience, they met their misfortune in a spirit that won admiration everywhere. Relief funds came generously from all quarters, both locally and abroad, and the Red Cross sent in large stores of supplies and clothing that proved invaluable. In the midst of widespread personal difficulties, it is pleasant to record that most of the old Far Eastern hongs considerately made good to their men the loss of their private effects and set them up afresh. The feudal system at its best.

With cold weather approaching, I made a quick trip to Shanghai to re-outfit and to arrange for the return of Dorothy and the children. James Thomson, in similar plight, was a fellow-passenger on the N.Y.K. express ship *Atsuta Maru*; and since I was engrossed in writing this narrative and the few other passengers were seasick or uncommunicative, he was driven to reading from cover to cover the only book in the ship's lounge, Bradshaw's railway guide to England, claiming at the end to be the only man in Christendom able to tell how to get from one point to another in the South of England without going up to London.

This brings me to the last pages of my original manuscript, which I shall transcribe as written:

'I began this account on my way up to Yokohama for the first time early in September. It is now November 11th and I am voyaging back to Kobe from Shanghai, having seen Dorothy and the boys and arranged for their rejoining me. Thank God they are well and, under the loving care of my sister and 'little Emily', the horror of the earthquake has nearly worn off. The children no longer start at any sudden sound nor cry out in their sleep. They will all come over in December and we shall live in Kobe, now head-quarters in Japan. By then our bachelor mess in the 'San-bon-matsu' house will have dispersed.

'Bateman and Anderson go back next week to open our new office in Tokyo. Barnett is to be stationed in Yokohama. Our Japanese staff will resume their former roles, the seniors at once, the others as quickly as feasible.

'Gladys Bateman and Mrs Syme Thomson have gone home to England; Honor and Patsy Anderson back to their native Australia. The E. A. G. Mays are slated to be transferred back to the China field in Shanghai. And so with our relatives and friends, Dorothy's parents, the Campbells, will return to Yokohama as soon as a home is available. Mabel Fraser has gone home to England for good, with Mrs Wheeler. The Standard Oil are installing Van Smith and Helen in the Imperial Hotel in Tokyo. Charles and Winnie Rice are to remain in Kobe for the present. I mention only these few to given an idea of the general pattern. But most of those who left Japan in the relief ships have since gone on to their homelands, many never to return. The old, intimate close-knit colony has scattered to the four corners of the earth. Even the familiar old place itself has gone with the wind; and all that has meant 'home' to so many through the years, all the ties of affection and treasured association, are now but memories.

'Yet it is magnificent how courageously those who are left have turned to the appalling task of building afresh. The spirit to fight back seems unquenchable. Man is indeed a tough creature and hard to vanquish!'

❧ VI ☙

AFTERMATH

People have asked me: 'What happened to Yokohama afterwards?'

It is not easy to draw a line and say 'Here the episode of the earthquake ended and here the aftermath began'; because, like a thunderstorm receding into the distance, the reverberations of calamity diminished slowly. For me, the line was drawn by circumstances. Twenty months after the earthquake, our Chairman wrote suggesting that as soon as I felt our affairs were reasonably restored to order, I should take my family over to British Columbia on three month's recuperative leave. We left Kobe in the *Empress of India* on July 7, 1925 and towards the end of our pleasant sojourn in Oak Bay, Victoria, I was asked to relieve an emergency by taking over our New York office. The transfer became permanent; I was later made the Company's Director in the United States and Canada, and never returned to Japan.

As to what ultimately happened to Yokohama, I can only speak from hearsay. Certain things started while I was still there but I did not witness their completion.

A dangerous threat to the restoration of Yokohama's former commericial importance was the impending loss to Kobe of the valuable silk trade. Foreign silk merchants—mostly Swiss, Italian and Continental, with their teams of experts had perforce quickly re-established themselves in Kobe, and the prized commodity was soon flowing smoothly through the southern port. Silk firms were prominent amongst those reluctant to contemplate rebuilding in Yokohama. But the City Fathers met this situation ingeniously, by offering to reconstruct destroyed premises to any desired specification, rent them to the silk firms at ten per cent of cost, and at the end of ten years turn the premises over free. It was a master-stroke. Up went the premises, back went the silk firms, and the valuable trade became once more a virtual monopoly of the northern port.

Awakening to the discouraging influence upon confidence of the continuing spectacle of ruin in the foreign Settlement, the Japanese authorities came up with another bright idea. If property owners would relinquish all right to the debris of their buildings, they would clear it down to ground level, dump it off the Bund into the harbour and make a park of the resultant reclamation. This timely offer was gladly accepted and the new park now extends about a hundred yards beyond the old sea-wall of the Bund. Laid out with trees and formal walks, it makes a pleasant promenade for strollers, though the Bund is no longer the open waterfront boulevard of early days.

Before permanent rebuilding on the old sites could render impossible any basic change, the municipal authorities, in collaboration with property owners, set about widening some of the narrower streets in the Settlement and opening up dead-end lanes to connect the main streets. To what extent this was carried out I do not know, but it was a very worthy project, as the streets of Yokohama were designed for rikishas, not automobiles; and in the Japanese city the need for wider thoroughfares was even more pressing.

Gradually new buildings were erected on the cleared lots, but not with the confidence of pioneer days, as it became evident that many concerns had decided to make Tokyo their headquarters and be content with lesser establishments in Yokohama. A tendency to get along with temporary quarters put the brakes on ambitious redevelopment, although a few handsome edifices here and there gave evidence that faith in the future was not altogether lacking.

It was some time, however, before a new Grand Hotel came into being on the site of the old at the corner where the Bund meets the Creek—a fine modern hostelry, but without the alluring front terrace so distinctive of the old. Even more sorrowful to the old-timer is the reclamation of the foreshore directly opposite the Grand, projecting the entrance to the Creek several hundred yards out into the harbour; and this has been built up with what appear to be facilities for warehousing ship's cargoes. So even if the famous Grand Hotel terrace had been resurrected, its delghtful panoramic view from the harbour down the Bay would no longer exist.

The Bluff, it is said, has greatly changed in character, much of its quaint picturesqueness having been lost in the transition to modern ranch type or concrete houses. But for familiar twists in the Bluff roads, an old resident has difficulty in placing himself. An entirely new colony of foreign residences has appeared on the hills surrounding the racecourse and out by the Cricket and Athletic Club grounds, overlooking 'Mississippi Bay'— the next large bight around the Honmoku cliffs below Yokohama, where the fishing villages of Negishi and Sugita fringe the shore. In the old days it was a lovely sight when the fishing fleet of forty or fifty sea-going sampans came gliding up the bay of a summer evening, their lofty oblong sails gleaming pink in the rays of the setting sun. The sound of their conches floated over the water, bringing out the villagers to assist in pulling the boats up the beach and unloading their heavy catches of fish. But the curse of reclamation has struck here too, and huge areas of tidal flats along the shore—the scene of many a happy clam-digging picnic—have been converted to geometric patterns of ochre-coloured building sites, already covered with drab industrial structures.

With a greatly diminished foreign population, the old social and sports clubs, hopefully resuscitated, had a hard time keeping their heads above water. Finally, several decided to merge, even the Yokohama United Club giving up its historic place on the Bund to link up with one of the Sports Clubs. But this has only come about in recent years.

In spite of its crippled condition, something of the old life did return to Yokohama and those who lived there through the thirties enjoyed it. Then came World War II and once more Yokohama was wiped out, this time by bombing. The Japanese city took the brunt of it, the Settlement and Bluff being largely spared. And yet again the port has come back to life, continuing its own special role. But this is a chapter in history that I must leave to others.

The Yokohama I still see in my mind's eye is the old one created by the pioneers, with its open bay and virgin hills I roamed in as a boy. Thoughts of those scenes and of the community I knew so well bring on a nostalgia that is like the scent of incense in a temple grove. In this I know that I am not alone.

INDEX OF PERSONS MENTIONED

Abbey, David, 109
  Tom, 80
Ackland, R. J. 'Paddy', 79
Anderson, Frank J. 30–38 *and constantly*
  Honor and Patsy, 30, 43, 64, 70, 72, 106, 112, 127
Allcock, George, 47
Arias, Henry, 91

Babbitt, Miss, 100
Barnett, John P., 30, 32, 34, 90, 91, 123, 127
Bateman, A. E., 30–36, 52–53, 64, 70–72, 115, 127
  Gladys, 52–53, 56, 64, 70–72, 112
Bell, Gordon W., 30, 90–91, 106
Blatch, Wm D., 102
Blundell, George, 87
Blum, Paul, 42
Boulter, R., 99, 102
Bowden, Rothwell, 61, 62, 88
Brady, G. G. 'Tim', 48
Brigel, Jos., 80
Brockhurst, G., 74
Brown, Percy B., 55, 113
  Edna, Norman and Michael, 113
Bugbird, F. M., 122

Caldwell, Jock, 100, 102, 122
Campbell, W. W., 'Commodore', 44 *and constantly*
  Mrs W. W. 'Calla', 44, 57, 70, 104, 110, 127
  Dorothy, II, 17 photo
Carew, H., 49
Caro, Don Jose, 65
Carst, Capt. J., 86
Clarke, R. M., 80

Colton, George W., 30, 90–91, 106
Cotte, L., 68
Coutts, Edward, 51, 64, 122
Cullinan, F. J. F., 80

Daimyo, 46, 70, 78, 92–97
Davies, Wm, 100
Dejardin, Paul, 65
Dentici, M., 63
Dodwell & Co. Ltd, 29
Duncan, Miss, 83

Edwards, R. C., 43, 73
Ewing, Jas. H., 111

Fardel, H. L., 55
Favre-Brandt, 36, 37
Feast, W. G., 68
Fitzgerald, M., 82
Fox, Eugene, 79
  Lucy, 30–32, 90–91, 106
Fraser, Mabel, 77, 82, 97, 112, 127
Frazar, E. W., 123
Fukui, 32

Geary, John R., 124
  Mrs John R., 63, 71
Gill, Robert, 80
Gillon, O. T., 89
Goodban, J. H. C., 48
Gripper, H. E., 68, 89, 91
Guiness, Alan, 100, 102–103

Hall, John W., 80
Haight, W., 99
Harris, Townsend, 17
Hayward, W., 88
Herr, 113
Heseltine, Capt. C., 102, 121

133

Hill, Fred. W., 41, 89, 114
  Winnie and Doris, 41
Hill, E. R. 'Bob', 115
Hilton-Johnson, Major, 112
Hingston, Surgeon Com., 51
Homewood, G., 79
Horne, Hugh, 99–100
  Grace and Itta, 99–100

Ichi, 69–70
Ishii, 42, 45

James, Capt. David, 113
Jensen, Torleif, 86
Jeffrey, E. C., 30–34, 88–91, 106
Jenks, Paul, 100
'Jimmy', 79

Kane, 44–46, 49–50
King, E. J., 47
  W. and Mrs, 44, 58, 70, 71
Kirjassoff, Max, 100
Koenings, Capt., 98
Koshimura, 35
Kruger, Ken, 97

Laffin, Tom, 93, 103–104
Lauritsen, Miss, 109
Lees, 99
Lefroy, A. J. S. 'Pat', 79
Lester, Hugh, 112
Loftus, Capt. E., 82
Lucas, Stephen 'Tiff', 93, 103–4

Macdowell, 110
Macgregor, M., 51
Mackenzie, D. J., 82
McKenlay, J. R., 63
Maitland, N. George, 105, 112
  Eleanor and Donald, 77, 105, 112
Manley, August, 82, 83

Martin, C. K. Marshall, 63, 124
Martin, Mary, 30, 32, 90–91, 106
Mason, W. B., 80
May, E. A. G., 111, 114–5, 127
  Doris, 112, 115
Melhuish, Ralph P., 112
Mendelson, Alice, 113
  Morris, 52, 81, 108
  Madeline, 52, 81–82
  general, 42
Milne, Prof., 96
Mine, 45, 50, 57, 107
Moilliet, George, 72–4, 95–6
Mollison, John, 99
Morris, 109
Morrison, I. C. 'Stinie', 73
Moss, R. F., 105
  E. J., 49
  Sir George, 112
Murray, Mrs May, 106

Nicoll, H. H., 73
Niven, G. S., 80

Owston, Capt. Francis, 87

Parkhouse, C. E. D., 51, 81
Paravicini, Dr, 96
Pattisson, Capt. P. B., 33 photo
Perry, Commodore Matthew C., 17
Poole, Otis A., II, 77, 105
  Herbert, A., 110, 112
  O. Manchester, II, *and throughout*
  Dorothy, 34, 42–46, 76, 105, 127
  Anthony, 29, 45–47, 57–58, 62
  Richard, 25, 45, 59–62
  David, 29, 45, 58–62

Rankin, Harry, 68
Reynell, Georgie, 113

Rice, Col. Elisha E., 17
  Charles R., 74, 82–3, 87
  Winnie, 73, 83, 127
  Ken and Nancy, 83
Robertson, A. 'Robbie', 79, 91
Robinson, Capt. Sam, R.N.R., 95
Rogers, Mrs Edward, 63
Rose, 89
Rowbottom, H. W., 54–55
  Flora, 55
Russell, Maurice, 52, 55, 64–65
  The Misses, 63

Sandberg, P., 90
Scott, A. P., 119
Sharp, Hugh P., 41
  Joan, 42
Shea, Frank, 73, 110
Simpson, Capt., 96
Smedley, Mrs, 88
Smith, N. Ferdinand, 85–87, 107
  K. van R., 86–7, 108, 110, 127
  Helen and Maureen, 86–7, 108, 110, 127
Spence, C. L., 113
Squire, Wm, 109
Stanton, Cyprian, 87
Stapleton, H. T., 60
Starr, 66

Stott, J. S., 31, 34, 82
Strong, Rev. Eustace, 77, 102
Syme Thomson, Mrs, 52–56, 64, 70–72, 106, 112, 127

Tait, Arthur and Hilare, 87
Taylor, H. W., 75
Tebbut, F. J. H., wife and child, 113
Thomson, Jas A., 30–34, 90–91, 122
Thorn, Charles, 108
  Muriel, Jack and David, 108
Tipler, Montague, 100
Tipple, Capt. Rennie, 68

Waddell, R., 99
Warming, Sophus, 83
  Fanny and Karin, 83
Warrener, 42
Watson, L. 'Jock', 42, 79, 119
Wheeler, Dr Edwin, 39, 48, 80, 112
  Mrs E., 105–6, 112, 127
White, A. T. 'Tiny', 44
Wright, Frank Lloyd, 119

Yarnell, D. E., 64